EVERYTHING ABOUT PLC PROGRAMMING

Practical lessons on PLC programming using Allen Bradley, Siemens, and Mitsubishi PLCs with examples

Avinash Malekar

Copyright © 2021 Avinash Malekar

All rights reserved

No part of this book may be reproduced, or stored in a retrieval system, or transmitted in any form or by any means, electronic, mechanical, photocopying, recording, or otherwise, without express written permission of the publisher.

ISBN: 9798787141566

Dedicated to…..

Learners

CONTENTS

Title Page

Copyright

Dedication

Preface

Weekly newsletter

Virtual training kit using easy PLC

1. Introduction	1
2. Allen Bradley PLC	22
2.1 RS Logix 500	39
Case study 1	74
2.2 Studio 5000 Logix designer	78
Case study 2	132
3. Siemens PLC	142
3.1 TIA Portal	148
Case study 3	209
4. Mitsubishi PLC	218
4.1 GX works 3	220
Case study 4	249
5. Remote PLC program access	255
Assignments	264
Books In This Series	269

PREFACE

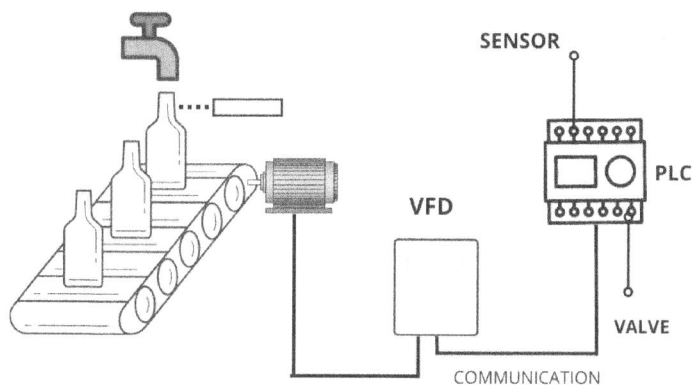

In the previous 'Everything about factory automation' book, we got introduced with the factory automation. We also came to know that a PLC is an unavoidable part of an industrial automation. An industry cannot be automated without the aid of a PLC.

There are so many PLC manufacturers available in the market, and they offer their PLCs with different aspects. Even though they are dissimilar, they follow the same working principle.

In this book, we will dig deeper into the basics and advanced PLC programming. It is not possible to cover all the PLC, so for the better comprehension I have chosen 3 popular PLC manufactures, i.e. Allen Bradley, Siemens and Mitsubishi. While studying we will focus not only on PLC but also other things that we need to know during writing a program. This book focuses on:

1. PLC- We will focus on introductory part of the different series of PLC that manufacturers offer with their suitable programming software. I have chosen 3 PLCs from 3 different manufacturers.

2. Communication- We will study about the different communication protocols normally used in automation.

The communication protocols are required to establish a communication between PLC and a PC.

3. Programming software- We will focus on the latest version of PLC programming software as per the different PLC manufacturers.

4. HMI- This book is only about the PLC programming, so we will cover the role of HMI in PLC programming.

5. VFD- This includes speed control, condition monitoring, and control commands of VFD through PLC programming.

6. Troubleshooting- We will cover the troubleshooting method you need to follow during breakdown.

The details of PLC and software we are going to use for study:

Click on this link for software assistance. PLC tutorials newsletter

Or scan this QR code

A. Allen Bradley (Rockwell Automation)

1. RS Logix 500

PLC- Micro Logix 1400
Application- Small size application

2. Studio 5000 Logix designer

PLC- ControlLogix and CompactLogix PLC
HMI - PanelView standard
VFD - Powerflex 525 EENET
Application- Medium/Large size application

B. Siemens

TIA Portal V16
PLC- S7 1200
PLC Programming-
HMI- KTP 700
HMI Programming- TIA portal V16
VFD- Sinamics V20
Application- Medium/Large size application

C. Mitsubishi

GX Works3
PLC- FX5U
HMI- GOT 2000
VFD- D700
Application- Small/ Medium size application

What makes this book different?

1. Well organized & digestible information

2. Program elaboration with latest version of programming software

3. Three popular PLCs in one book

4. HMI & VFD programming

WEEKLY NEWSLETTER

Discover the cutting-edge world of industry automation and stay ahead with PLCTutorials.beehiiv.com! Our weekly newsletter is your gateway to the latest technologies like IoT and IIoT revolutionizing industrial landscapes.

Don't miss out on this golden opportunity to level up your automation game. Subscribe today and be the first to harness the power of innovation.

PLC Tutorials

Dive into a wealth of knowledge:

- Access software downloading guides for seamless automation integration.
- Explore real-world automation case studies for practical insights.
- Stay updated with the freshest technology news impacting your industry.

Subscribe our weely newsletter by clicking on the following link.

PLC tutorials newsletter

Or scan this QR code

VIRTUAL TRAINING KIT USING EASY PLC

Hey there, fellow learner! Are you interested in PLC programming, but feeling discouraged by the high cost of physical PLCs? Well, have no fear, because I have some great news for you!

Gone are the days when you had to rely solely on training kits to test your PLC programs. These days, there are more affordable and accessible options available, such as the Easy PLC training suite.

Not only is the Easy PLC more budget-friendly than traditional physical PLCs, but it also provides a user-friendly interface that's perfect for beginners. With its intuitive design and comprehensive tutorials, the Easy PLC makes it easy to learn and master PLC programming.

So why spend a fortune on expensive physical PLCs when you can achieve the same level of proficiency with a more affordable option? Give the Easy PLC a try and see for yourself just how accessible and easy-to-use PLC programming can be!

Easy PLC suit

Easy PLC software suit is a complete PLC, HMI, and Machine simulator software package. It includes:

1. Easy PLC - PLC simulation will allow programming in ladder, grafcet, logic blocks, or script. It has a virtual PLC you can connect with it virtually.

The programming software has all the instructions you need for programming.

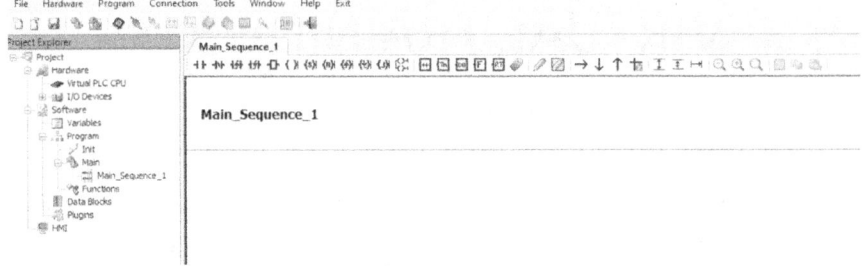

2. HMI system - Easily create a virtual HMI that no any software offer.

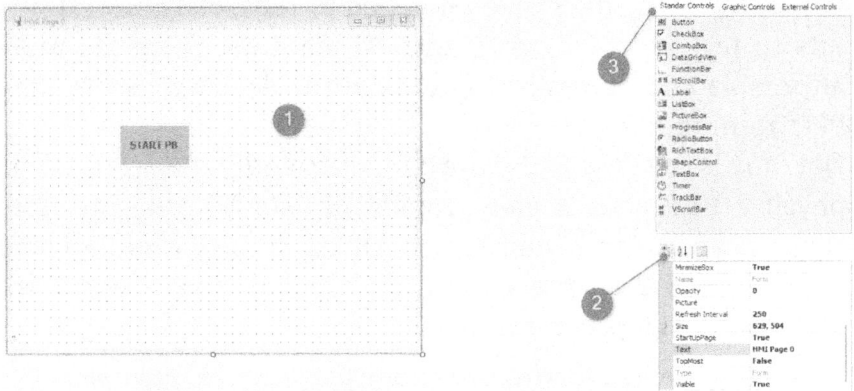

1. HMI screen
2. Properties

You can change the fort size, color of text and set connections of the devices.

3. Tool Box

a. Standard control

2. Graphics control

You can find different indicators such as level, LED, numeric, etc.

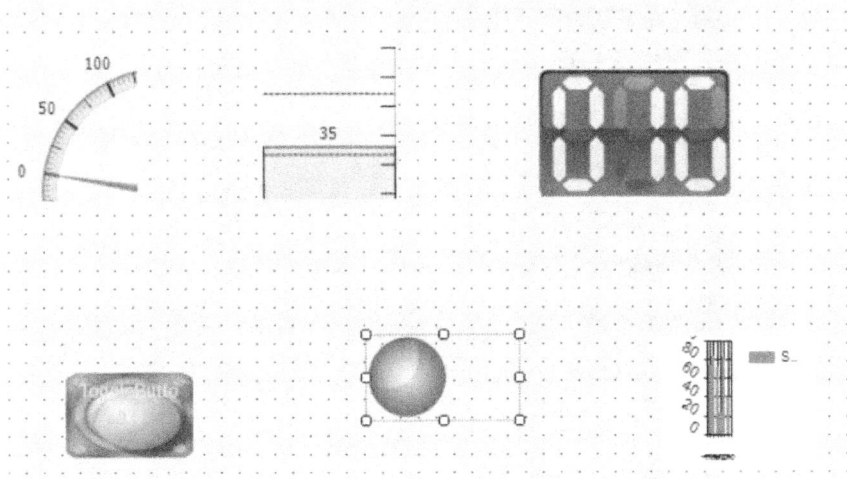

3. External control

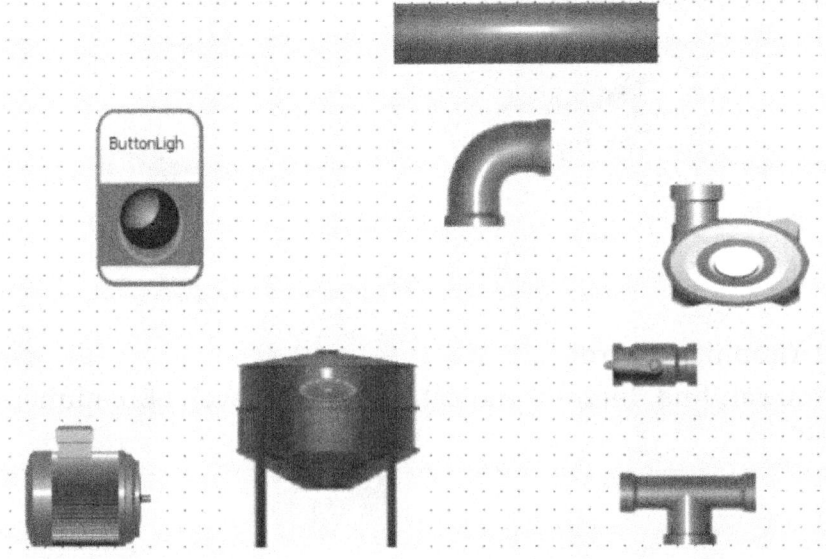

3. Machine simulator - A virtual 3D world with real time graphics and physical properties. PLC program can be tested using EasyPLC or through other interfaces like Modbus RTU, TCP, etc.

a. Library

The machine simulator provides you with the library of machine. You can use develop your programming skills using these machines. These machines have demo mode so that you can understand the working of the machine.

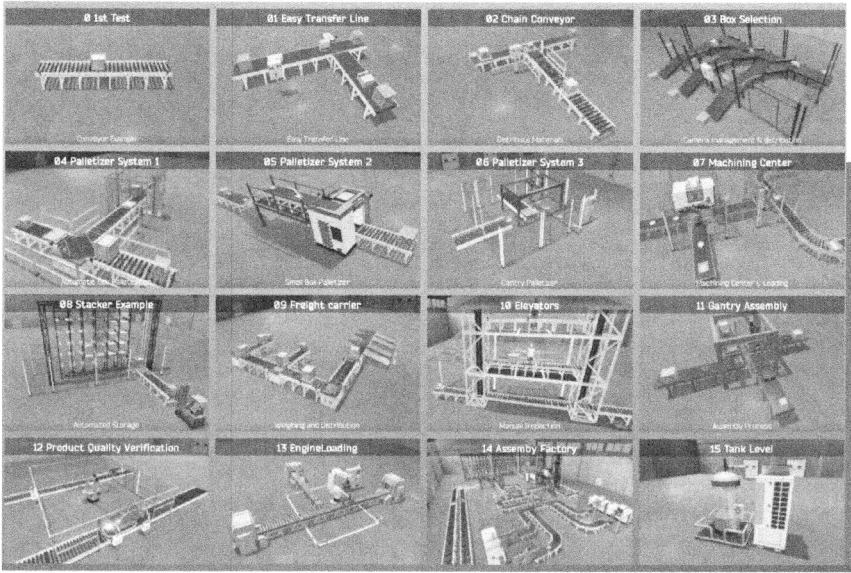

b. Editor

The machine simulator has also editor mode so that you can create you own machine set up by using different tools such as robot, conveyor, pneumatic cylinder, etc.

c. Machine simulator architecture

1. Vision option- You can different camera options such as drone, 3D camera, human to view the working of your machine.

2. PLC control- You can control the PLC operation from here.

3. IO Drivers

4. Outputs- All the digital and analog output devices of the machine is shown here with their status.

5. Inputs- All the digital and analog input devices of the machine is shown here with their status.

Install Easy PLC suit

1. Download Easy PLC demo

Visit http://www.nirtec.com/index.php/plc-software-download/

Or scan the QR given above.

2. Select software version

3. Extract and run as administrator

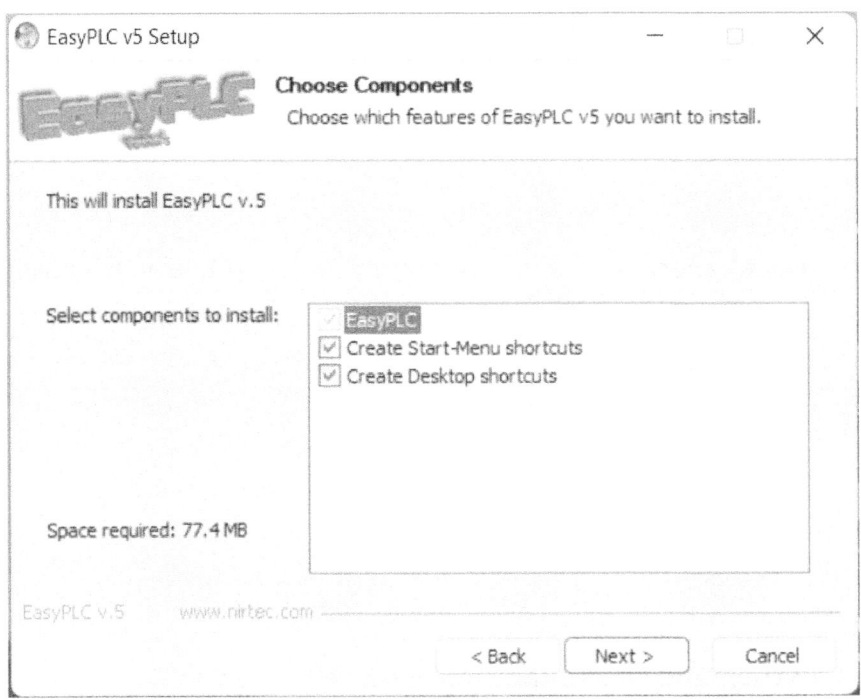

After installation, you can see these three icons on your desktop. Now you need machine simulator which is not available for free. Please follow these steps to purchase the MS software.

4. Steps to purchase Machine simulator

Visit https://www.nirtec.com/purchase-price/

Easy PLC products:

   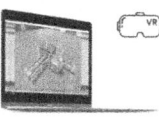

Standalone License
85,00 €

USB License
95,00 €

Tutorial Manager
60,00 €

Machines Simulator VR
60,00 €

USB License + Interface Card
280,00 €

a. Standalone license- The standalone license allows EasyPLC + Machines Simulator to access a single computer. This type of license can be activated/deactivated as many times as necessary, allowing it to be moved between computers.

You can only run the software on one computer at a time.
IMPORTANT: With this type of license it is necessary to have an Internet connection while using the software.

Using this license you will have access to:

- Machines Simulator & Machines Simulator Lite

- EasyPLC + Virtual PLC + HMI System

This license does not expire over time nor does it nedd to be renewed.

b. USB license
All software modules are installed directly from the USB key. The USB license gives you the ability to have the software on multiple operating systems for computers, laptops, or virtual computers but requires the USB stick to be plugged into the system that is being used. It is a single-use at a time but multiple uses from access to access if you change computers.

Internet connection is not required using this license!

Licenses do not expire over time nor do they need to be renewed.
Using this license you will have access to:
- Machines Simulator & Machines Simulator Lite

- EasyPLC + Virtual PLC + HMI System

c. Tutorial Manager

Tutorial package offers 20 practical automation examples with problem & solution about PLC programming. All exercises are solved for three different PLC systems: EasyPLC, Siemens TIA Portal and Rockwell Automation Studio 5000 Logix Designer.

This module will be added to the license you have choose (Standalone or USB)

d. Machines simulator VR

Do you want a more real experience? use Machines Simulator VR, the software designed for Virtual Reality!

Once you have created your system with Machines Simulator Editor (or using the already predefined systems), launch Machines Simulator VR, put your headset and enter in the REAL VIRTUAL WORLD, touching the environment, sensors, pallets, boxes, etc... You will test the virtual system like you were in the real world, a really amazing experience.

Machines Simulator VR introduces you to virtual reality through a VR headset compatible with Steam VR

e. USB license+ Interface card

The USB license gives you the ability to have the software on multiple operating systems for computers, laptops, or virtual computers but requires the USB stick to be plugged into the system that is being used. It is a single-use at a time but multiple uses from access to access if you change computers.

Using this license you will have access to:

- Machines Simulator & Machines Simulator Lite

- Machines Simulator VR

- EasyPLC + Virtual PLC + HMI System

- USB-OPTO_RLY88 Interface card

Internet connection is not required using this license!
The interface board is a way to wire external I/O directly to your PC to work with EasyPLC software if real-world I/O points are needed in your system.

Card specifications: https://www.nirtec.com/usb-opto-rly88-specifications/

Licenses do not expire over time nor do they need to be renewed.

GET UPTO 10% DISCOUNT ON ALL EASY PLC PRODUCTS ON USING MY COUPON CODE - Avinash

Installation of a standalone license

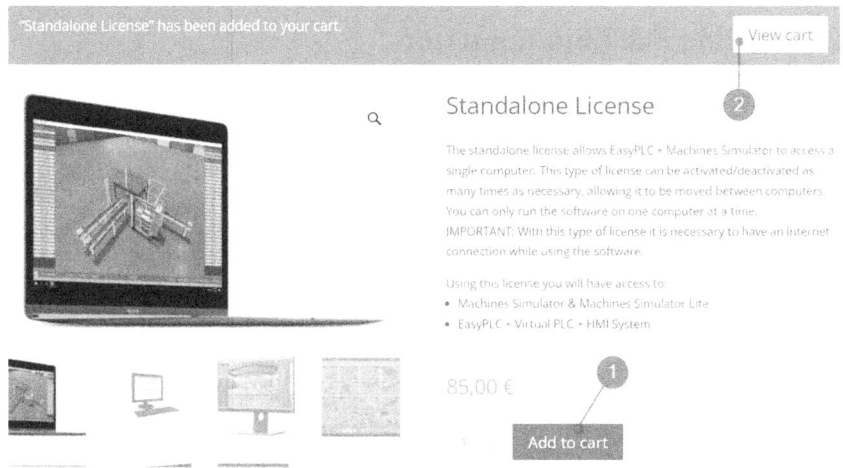

1. Select standalone license
2. Click on add to cart and then click on view cart.
3. Apply coupon code

Use my coupon code **Avinash** to get 10% off

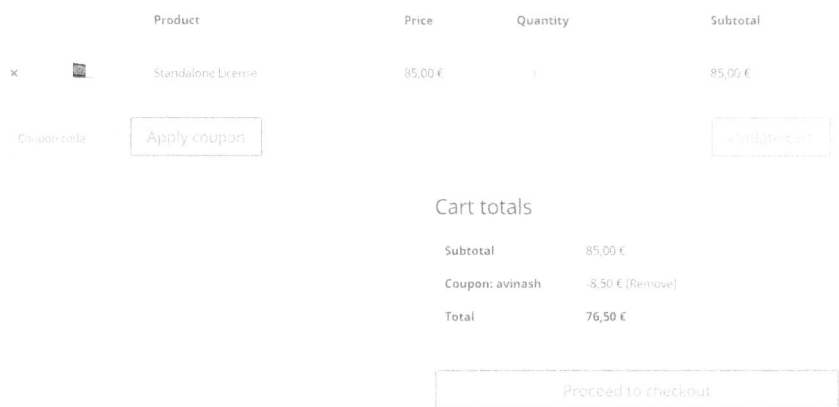

After applying my coupon code, you can see the purchase has reduced to **76.50 Euros.** Here you can select paypal or credit/

debit card methods to purchase this package.

Note:
- You get lifetime license at this cost, also you can update software version whenever it is available.

- You can use this PC license for only one computer.

5. Installing machine simulator

Once you complete the purchase process, you will have to share your PLC software id at easyplc@nirtec.com to get license.

Easy PLC> Help > About

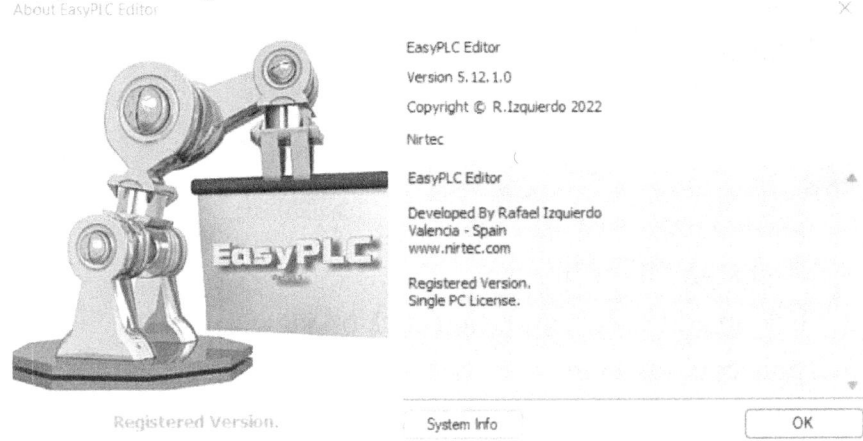

After that you will recieve an email from Nirtec team with the links to the rest of the Easy PLC software package suite and your license file. Install the machine simulation software.

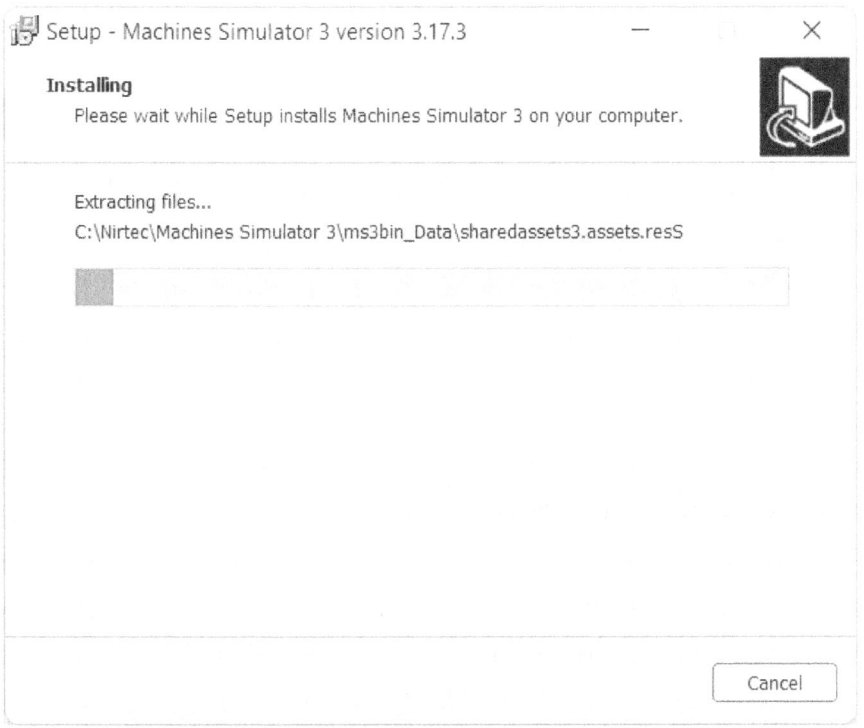

After completion of downloading process you can see one more shortcut on you desktop.

Now, copy the license file from the mail and paste it at directory of Easy PLC and Machine Simulator.

> This PC > Acer (C:) > Nirtec > EasyPLC v5 >

Name	Date modified	Type	Size
license	25-07-2014 19:37	Text Document	7 KB
log4net.dll	14-04-2009 13:30	Application extens...	264 KB
logo	14-07-2010 13:19	Icon	8 KB
Modbus.dll	22-08-2012 14:54	Application extens...	72 KB
ModbusTCP.dll	14-06-2013 19:01	Application extens...	12 KB
MotorHMI.dll	07-10-2015 00:48	Application extens...	1,127 KB
Nirtec.CodeEditor.dll	20-09-2011 17:16	Application extens...	400 KB
Nirtec.Docking.dll	06-10-2015 19:46	Application extens...	416 KB

Create a project using Easy PLC

In this section, I am going to show you the procedure of creating your own project using Easy PLC suit.

1. Create a project using machine simulator 3

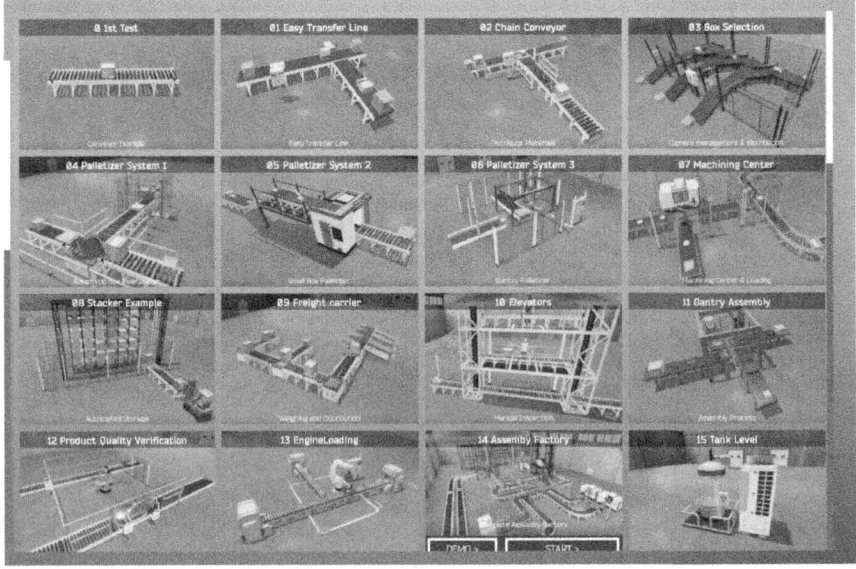

Once you open the project, you can see its IO devices also.

Open Easy PLC software and select the programming mode. I have selected machine simulated program from the list.

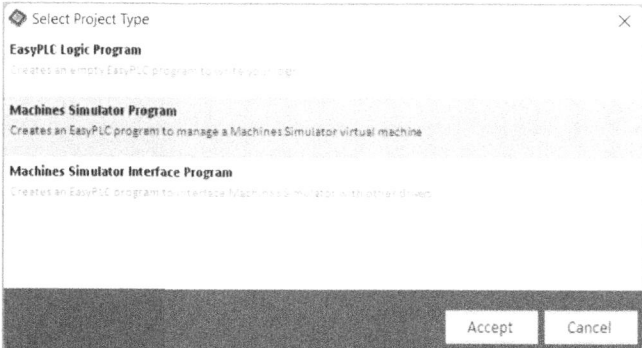

2. PLC program

a. You can select the program language from the list.

XXVII

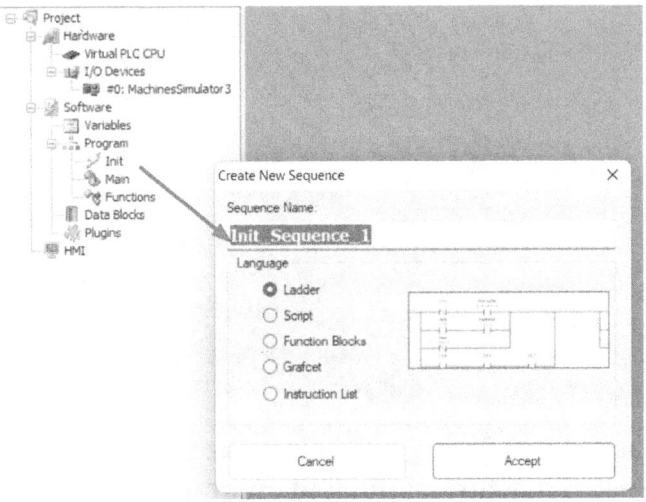

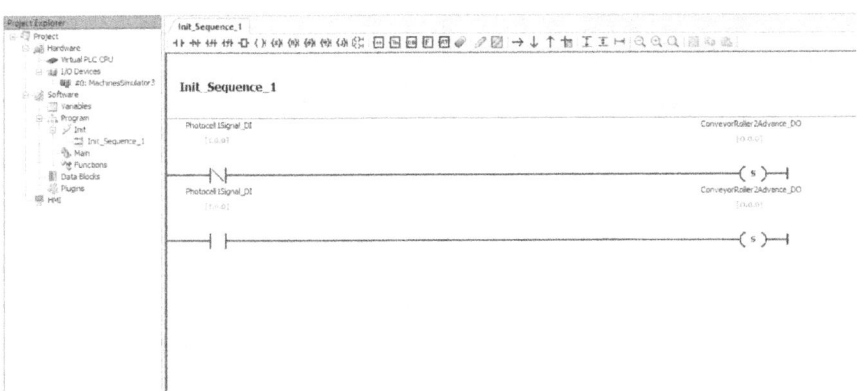

b. Go to Program and click on Compile and transfer.

c. Launch virtual PLC

d. Connect online with PLC

3. HMI Program

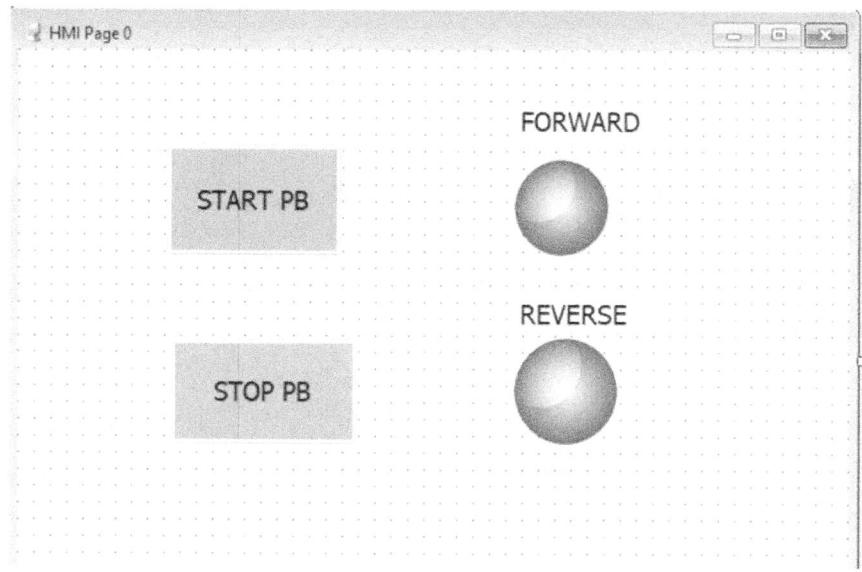

a. Right click on the HMI in Easy PLC software and select 'New HMI Page'.

b. Create two push buttons and LEDs, then go to properties and them tags.

c. Go to the connection option and select 'Launch HMI system' to run the HMI screen. Now you can control the machine from HMI.

4. Run Simulation

Open you project in Machine Simulator and start the PLC. Now you can see the working of your project.

1. INTRODUCTION

A. Basics of PLC

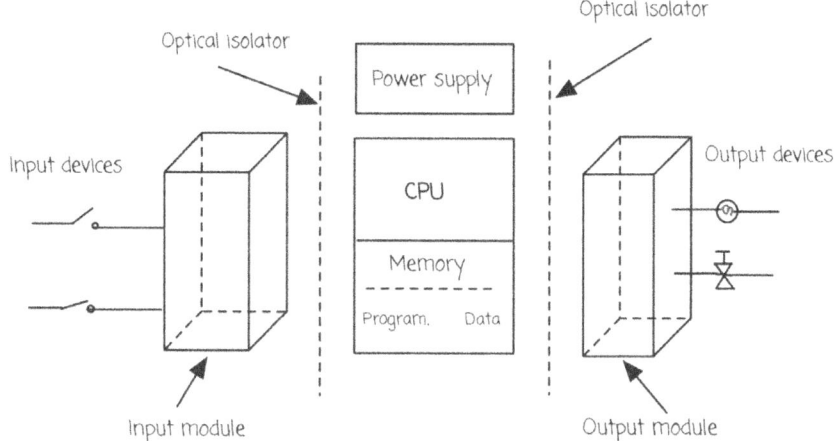

According to the National Electrical Manufacturers Association (NEMA), a programmable logic controller (PLC) is a digital electronic device that employs a programmable memory to store instructions and carry out particular activities like logic, sequencing, timing, counting, and arithmetic operations to manage machinery and processes. The PLC is a collection of solid-state digital logic components intended for logical decision-making and output generation. The interface between input sensors and output devices is preprogrammed.

The entire PLC system can be divided as follows.

a) Central Processing Unit
b) Input Modules
c) Output Modules
d) Power Supply
e) Bus system

a) Central Processing Unit

CPU is a brain of the whole PLC is the CPU module. This

module typically lives in the slot beside the power supply. The manufacturers offer different types of CPUs based on the complexity needed for the system.

The CPU consists of a microprocessor, memory chip, and other integrated circuits to control logic, monitoring, and communications. The CPU has different operating modes.

In programming mode, it accepts the downloaded logic from a PC. The CPU is then placed in the run mode so that it can execute the program and operate the process.

b) Input Module

These modules act as an interface between the real-time status of the process variable and the CPU.

	Selector switch		Through beam sensor
	Limit switch		Inductive sensor
	Push button		Limit Switch

Analog input module: Typical input to these modules is 4-20 mA, 0-10 V Ex: Pressure, Flow, Level Tx, RTD (Ohm), Thermocouple (mV).

Digital input module: Typical input to these modules is 24 V DC, 115 V AC, 230 V AC.

c) Output module

	Valve		Buzzer
	Lamp		Relay
	Solenoid coil		Motor starter
	Alarm		Fan

These modules act as a link between the CPU and the output devices in the field.

Analog output module: Typical output from these modules is 4-20 mA, 0-10 V

Ex: Control Valve, Speed, Vibration

Digital output module: Typical output from these modules is 24 V DC.

d) Power Supply

The power supply gives the voltage required for the electronics module (I/O module, CPU module memory unit) of the PLC from the line supply.

The power supply provides the isolation necessary to protect the solid-state devices from the highest voltage line spikes. As I/O is expanded, some PLC may require additional power supplies in order to maintain proper power levels.

e) Bus System

It is the path for the transmission of the signal between the power supply module, CPU, and I/O modules. The bus consists of several single lines i.e. wires or tracks.

- Data bus
- Address bus
- Control bus

B. Processor Memory Organization

The memory of a PLC is organized by type.
The memory space can be divided into two broad categories:
1. Program Memory
2. Data memory.

The advanced ladder logic functions allow controllers to perform calculations, make decisions, and do other complex tasks. Timers and counters are examples of ladder logic functions. They are more complex than basic input contacts and output coils and rely heavily upon data stored in the memory of the PLC.

A memory map can be used to show how memory is organized in a PLC.

1. Data table
-Input/output locations
-Internal relay and timer/counter locations

2. User program
The user program causes the controller to operate in a particular manner.

3. Housekeeping memory
Used to carry out functions needed to make the processor operate (no access by user).

C. PLC manufacturers and their market share

According to research by grand view research, the global industrial automation, and control system market was worth almost 126.3 billion USD and expected to rise 8.6% up to 2025. Companies are significantly reducing labor and operational expenses, and additionally minimizing human errors by the use of industrial robots through automation.

The manufacturers are using various control systems for their industrial processes. The primary control systems are Programmable Logic Controller (PLC), Supervisory Control And Data Acquisition (SCADA), Distributed Control System (DCS), and Human Machine Interface (HMI).

Japan
- Omron Industrial Automation
- Yaskawa Electric Corporation
- Mitsubishi Electric
- Fuji Electric
- Yokogawa Electric
- Toshiba
- Panasonic
- Keyence

United States
- Emerson Electric
- Rockwell Automation

Germany
- Siemens
- Bosch
- Phoenix Contact

Taiwan
- Delta Electronics
- Fatek

China
- Wecon Technology
- Kinco

Switzerland
- ABB

France
- Schneider Electric

Israel
- Unitronics

India
- RS Enterprises
- General Industrial Controls (GIC)

Based on the region, the market for industrial automation has been categorized into North America, Europe, Asia Pacific, Latin America, and the Middle East and Africa. A large portion of the growth is expected to come from the Asia Pacific. Kawasaki Robotics, Mitsubishi Electric Factory Automation, Yokogawa Electric Corporation are some of the key players from the Asia Pacific.

D. Types of PLC

1. Compact PLC

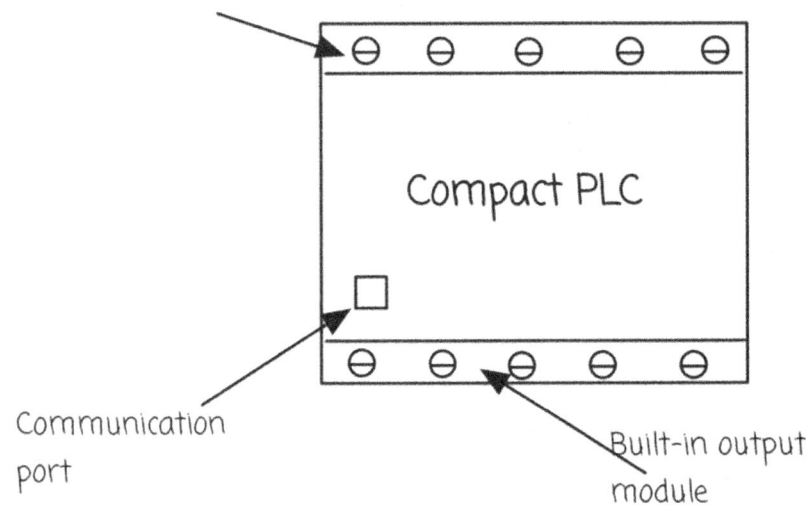

a. Compact PLC

This type of PLC comes with inbuilt input and output modules. It has a single port for communication to upload and download a program.

Application: Suitable for very small size machine application.

2. Compact PLC with expandable module

The compact PLC comes with a slot to mount modules for digital input and outputs, different communication modules (RS 232), etc.

```
              Built-in input module
                        ↓                 Expandable
              ⊖  ⊖  ⊖  ⊖  ⊖  ⏀⏀           module
                                ⏀⏀      ↙
                 PLC with       ⏀⏀
                 Expandable     ⏀⏀
                 chassis        ⏀⏀
              □                 ⏀⏀
              ⊖  ⊖  ⊖  ⊖  ⊖
Communication
port                    ↑
               Built-in output module
```

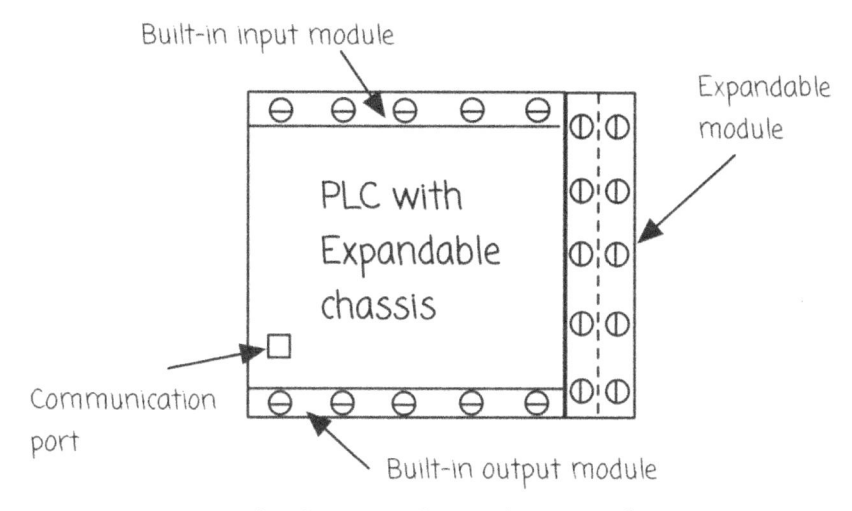

b. PLC with Expandable module

PLC: Allen Bradley MicroLogix series, Mitsubishi FX series, etc.

Application: Suitable for small and medium size machine application.

3. Modular PLC

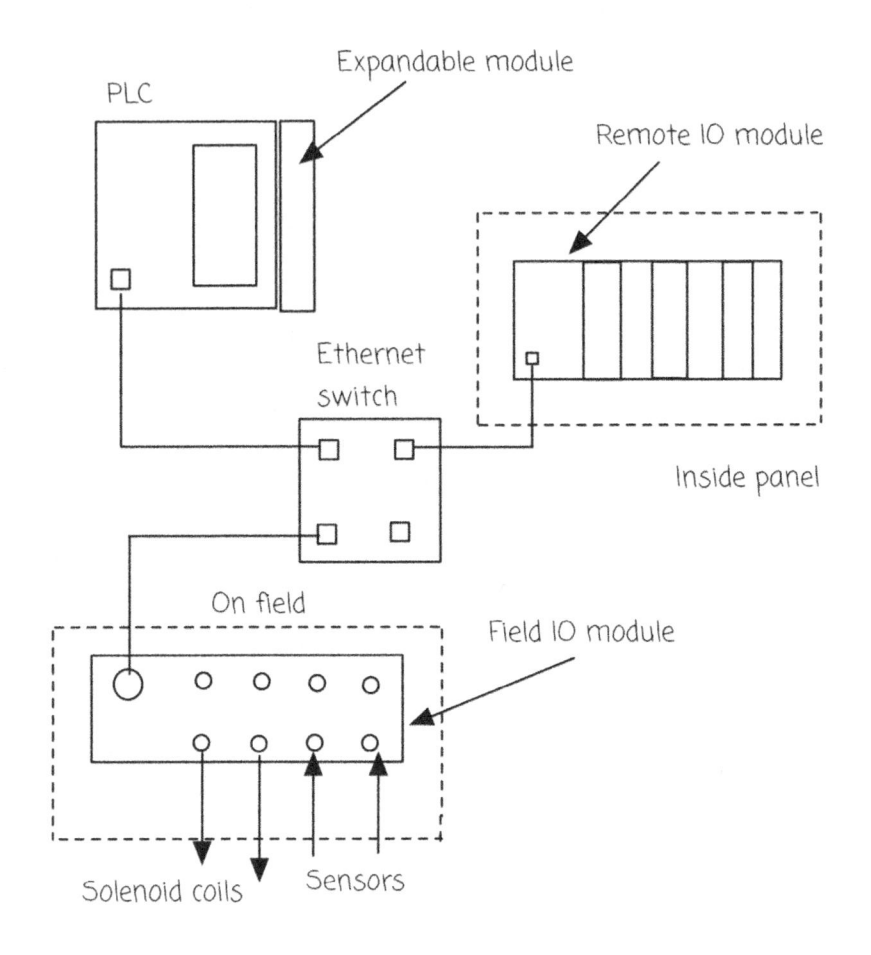

c. Modular PLC

This type of PLC is wide is used nowadays because of its performance and flexibility.

PLC: Allen Bradley CompactLogix and ControlLogix series

Application: Suitable for Medium and large size machine application.

4. Safety PLC

PLC with safety control

Safety input card

Remote IO

Safety output card

Safety input devices

Emergency pushbutton

Safety door

Safety light curtain

Safety output devices

Safety Relay

Safety contactor

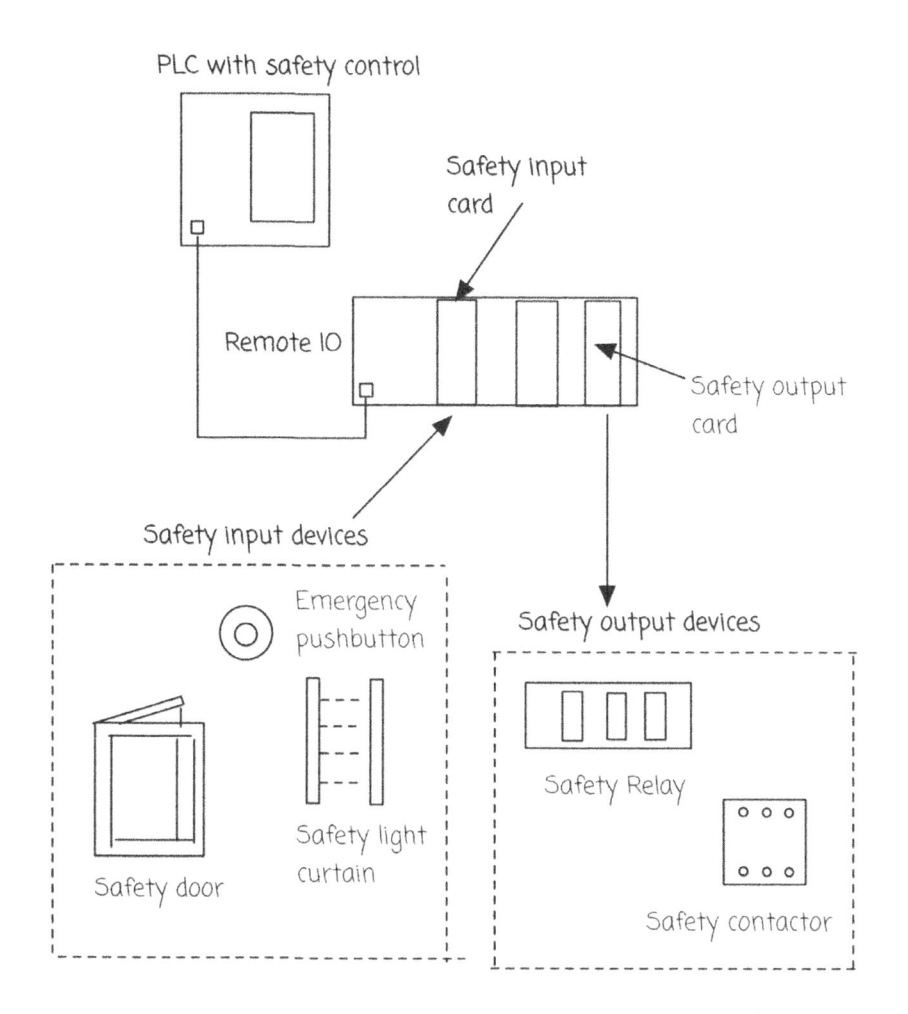

d. PLC with safety control

Safety is paramount in industrial settings. RSLogix 500 lets you design fail-safe mechanisms that halt operations in critical situations. Emergency shutdowns, door interlocks, and emergency stops become orchestrated dances with this software.

The controllers have the special features to provide the safety to the operator, they are especially designed for the safety so that you can connect different safety devices such as emergency switch, breaker, safety mat, light curtain, etc. to the controllers. and control the safety using safety contactors and relays.

PLC:

1. CompactGuardLogix
2. ControlGuardLogix series

Application:

1. Palletizer
2. Robot welding cell
3. Machining cell

5. Motion control PLC

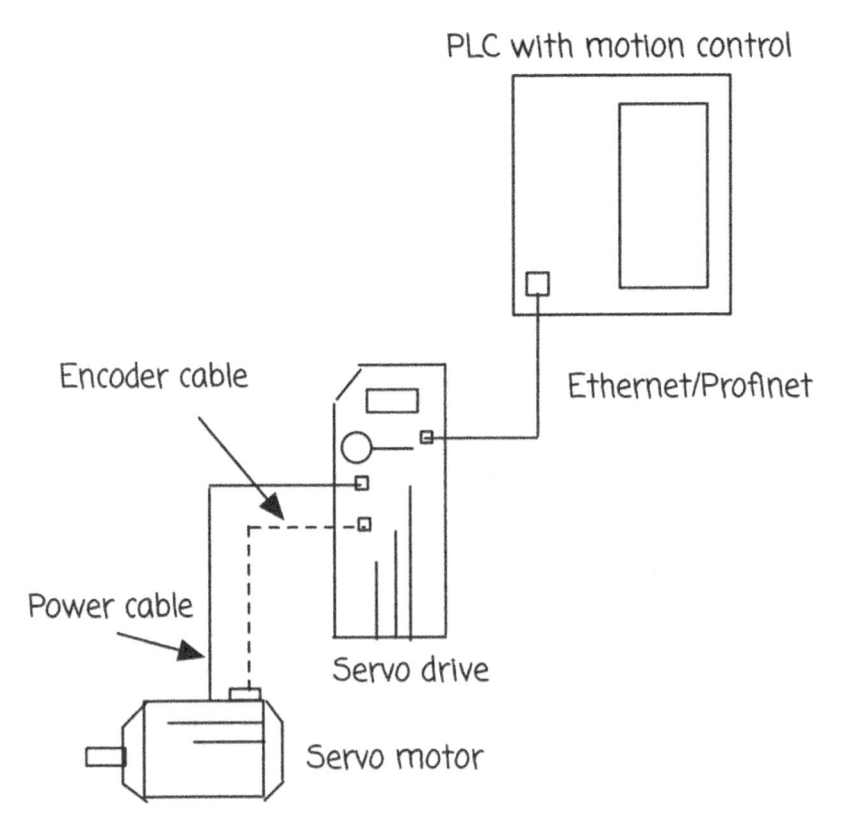

e. PLC with motion control

The PLC has special features for motion control. The PLC

communicates with Servo drive via communication protocol.

PLC: ControlLogix and CompactLogix ERM series with Kinetix servo drive.

Application:

1. Primary packaging machine.

2. Assembly of precise components.

E. PLC program execution

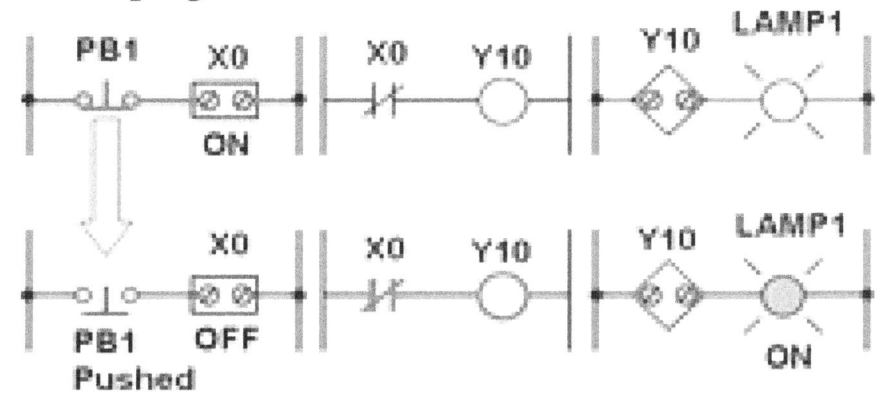

The below circuit shows the physical wiring along with the programming rung of the wiring.

Program execution process

a) Input Table File Operation

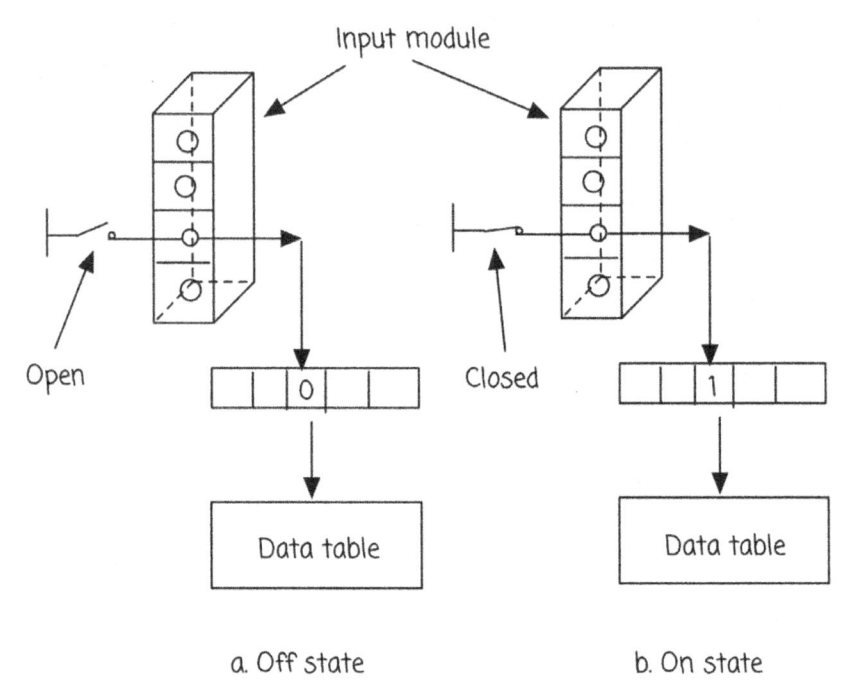

Input module

Open   Closed

Data table   Data table

a. Off state   b. On state

When the switch is open, 0 is stored in the file. Once the switch is closed, 1 is stored in the file.

b) Output Table File Operation

a. Off state   b. On state

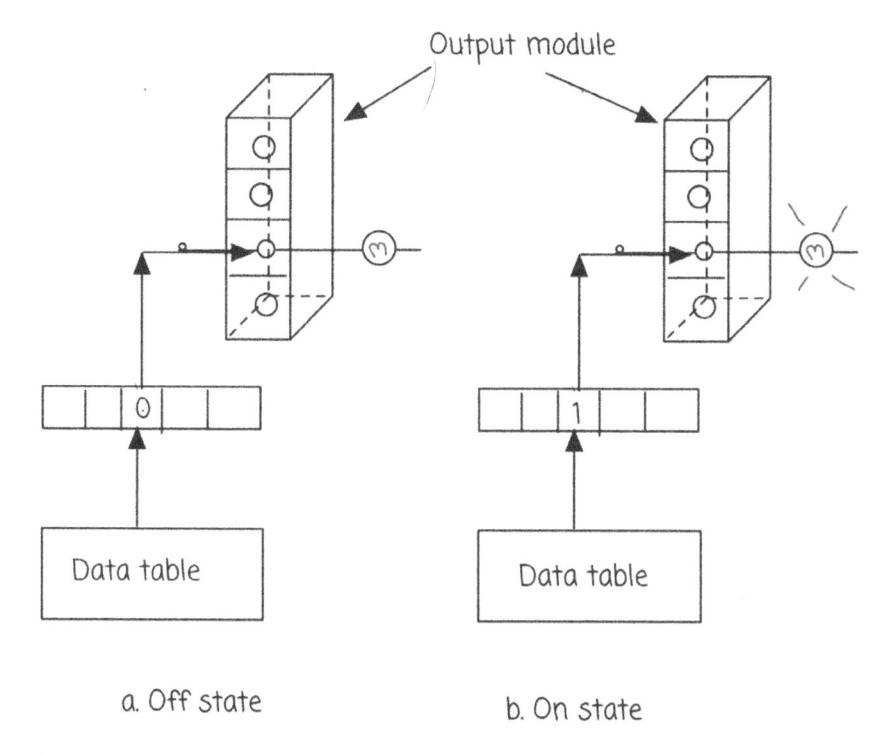

Since 0 is stored in the file, a lamp will not glow. Once 0 becomes 1, the lamp will glow.

F. PLC programming languages

PLC programming languages are defined by the international electrotechnical commission (IEC) under sections 61131-3 standards are explained below.

1. Ladder logic (LD)

A graphic language using ladder programs composed of contacts and coils. The inline structured text functions to edit ST programs on the ladder editor can be used.

```
    PB1   PB2            LAMP
  ──┤ ├───┤ ├──────────────O──────
    M00   │
  ──┤ ├───┘
```

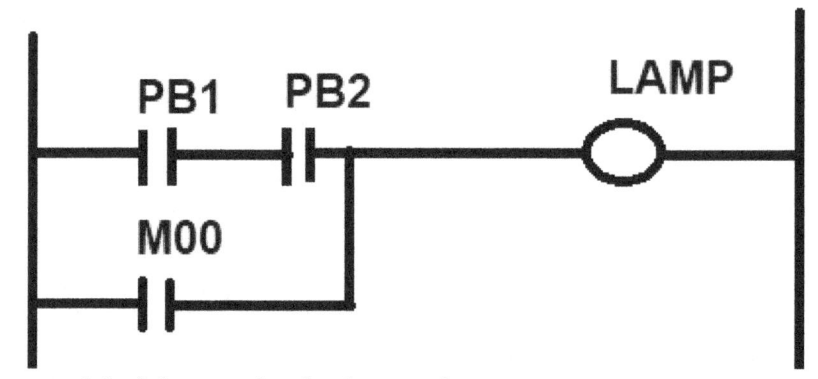

1. Modeled from relay logic panel.
2. Easy to understand and troubleshoot.
3. Difficult for motion and batching programming

2. Sequential function chart (SFC)

A graphic language for clarifying the execution order and the execution condition of a program.

1. Language is similar to flow charts.
2. The process can be broken into steps.
3. To have direct access to the logic for a piece of equipment fault.
4. Minimal programming effort and greater clarity through graphic programming.

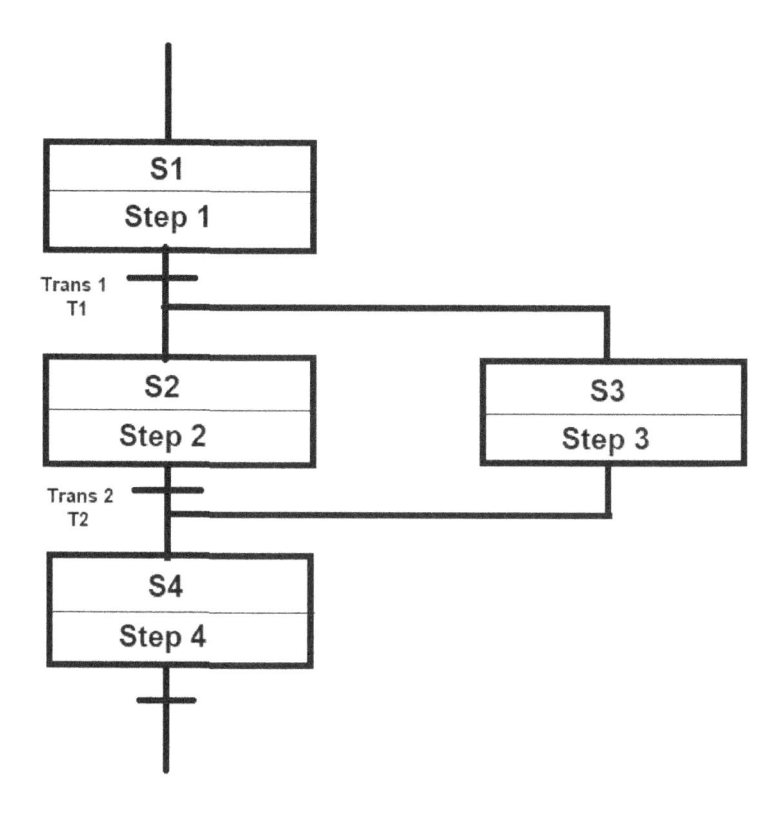

```
        ┌─────────┐
        │   S1    │
        │ Step 1  │
        └────┬────┘
Trans 1 ─────┼─────────────┐
  T1         │             │
        ┌────┴────┐   ┌────┴────┐
        │   S2    │   │   S3    │
        │ Step 2  │   │ Step 3  │
        └────┬────┘   └────┬────┘
Trans 2 ─────┼─────────────┘
  T2         │
        ┌────┴────┐
        │   S4    │
        │ Step 4  │
        └────┬────┘
             │
```

5. Excellent legibility for maintenance personnel.

6. Less time in the commissioning phase thanks to the graphical programming interface.

7. Minimal implementation effort because there are few possibilities for errors when generating code.

8. High availability of the machine through process diagnostics functions (interlock and supervision).

9. Fast error detection through PLC code display and criteria analysis on the HMI.

3. Functional block diagram (FBD)

A graphic language for creating a control program only by placing and connecting an element.

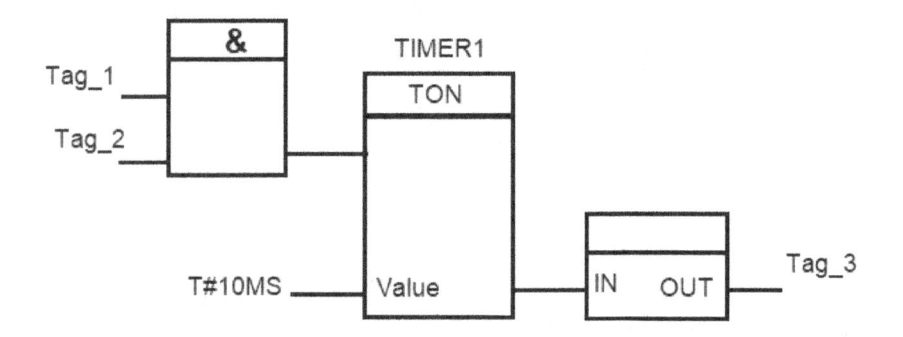

1. FBD describes functions between inputs and outputs that are connected by connection lines.
2. It is in the graphical form language with repeated blocks.
3. It is good for motion control programming.
4. It can combine many lines of the program into a single block.
5. But it is difficult to troubleshoot using this language.

4. Structured text (ST)

```
IF #IN1 & #IN2,
THEN
#OUT1 :=TRUE;
END IF;
"IEC_Timer_0_DB".TON (IN:=10,
                     PT:=15,
                     Q=>10,
                     ET=>10):
```

Control syntax such as selection branch by conditional syntax or repetitions by iterative syntax can be controlled, as in the high-level language such as C language. By using these syntax, concise programs can be written.

1. Similar to C language, this language uses instructions like- for, while, if, else, case, if-else, etc.

2. Very organized and good for computing large mathematical calculations.

3. It covers some instructions that are not available in the ladder diagram.

4. It is difficult to edit online.

5. Instruction List (IL)

An instruction list (IL) consists of a series of instructions. Each instruction begins in a new line and contains an operator and, depending on the type of operation, one or more operands separated by commas. In front of an instruction there can be an identification mark (label) followed by a colon (:).

A comment must be the last element in a line. Empty lines can be inserted between instructions.

```
LD      TRUE       (*load TRUE in the accumulator*)
ANDN    BOOL1      (*execute AND with the negated value of the BOOL1 variable*)
JMPC    label      (*if the result was TRUE, then jump to the label "label"*)
LDN     BOOL2      (*save the negated value of *)
ST      ERG        (*BOOL2 in ERG*)
label:

LD      BOOL2      (*save the value of *)
ST      ERG        (*BOOL2 in ERG*)
```

Modifiers and operators in IL

In the IL language the following operators and modifiers can be used.

Modifiers:

- C with JMP, CAL, RET: The instruction is only then executed if the result of the preceding expression is TRUE.

- N with JMPC, CALC, RETC: The instruction is only then executed if the result of the preceding expression is FALSE.

- N otherwise: Negation of the operand (not of the

accumulator)

Below you find a table of all operators in IL with their possible modifiers and the relevant meaning:

Operator	Modifiers	Meaning
LD	N	Make current result equal to the operand
ST	N	Save current result at the position of the Operand
S		Put the Boolean operand exactly at TRUE if the current result is TRUE
R		Put the Boolean operand exactly at FALSE if the current result is TRUE
AND	N, (	Bitwise AND
OR	N, (	Bitwise OR
XOR	(	Bitwise exklusive OR
ADD	(	Addition
SUB	(	Subtraction
MUL	(	Multiplication
DIV	(	Division
GT	(	>
GE	(	>=
EQ	(	=
NE	(	<>
LE	(	<=
LT	(	<
JMP	CN	Jump to label
CAL	CN	call function block

RET	CN	Return from call of a function block
)		Evaluate deferred operation

1. Instruction list language consists of predefined instructions, which are used to program a PLC.

2. The use of instructions makes this program very compact.

3. It is difficult to edit online.

H. PLC programming steps

1. Select a PLC
The first step of PLC program is to select a PLC from the list and set its IP address.

2. Select PLC programming language
When you open the software which is compactible to the selected PLC, select the programming languages from the list. You can use multiple language in a single program, provided their block are different.

3. Module configuration
Depending upon the number of IO devices, you can add modules with your PLC.

4. Configuration of peripheral devices
The devices like HMI, VFD, remote IO modules are configured with their IP addresses.

5. IO mapping
It is one of the crucial step of PLC programming. The IO mapping is done to make a program more informative and easy to understand.

6. Program writing
Rather than writing a program into a single block, it can be broken into small blocks such as manual, auto, alarm, etc.
It is necessary to call these blocks using call instruction in main folder.

7. Alarms

It is always better to create all the possible alarms, this will be benificial for operator for troubleshooting.

H. Data collection and monitoring with PLC

The machine is no longer limited to the production nowadays, it is capable to transmit the information to the server using latest technologies. There are many technologies available in the market that helps to connect a PLC with the server. So the customer can fetch the data from the server and use it in his ERP system.

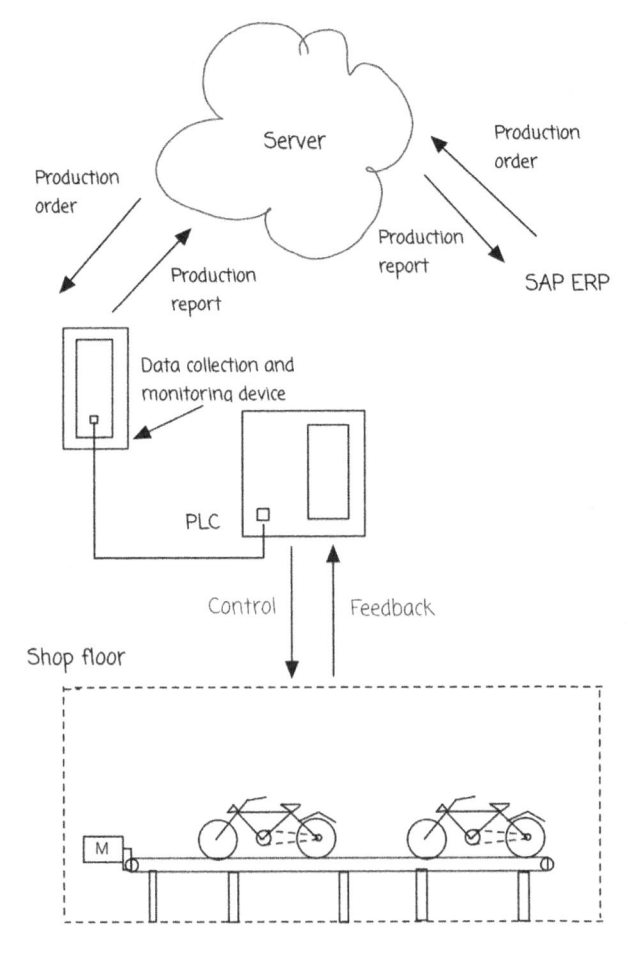

f. PLC with data collection and monitoring device

2. ALLEN BRADLEY PLC

Allen-Bradley became involved with programmable logic controllers by the inventor, Odo Josef Struger. He is often referred to as the father of the programmable logic controller. Odo contributed the ideas leading up to the invention of Allen-Bradley PLCs over a period from 1958 to 1960 and is accredited with inventing the acronym PLC.

Allen-Bradley became a major PLC manufacturer in the United States during his employment with AB. He also contributed as a leader in the development of IEC 61131-3 PLC programming language standards.

A. Allen Bradley PLC series

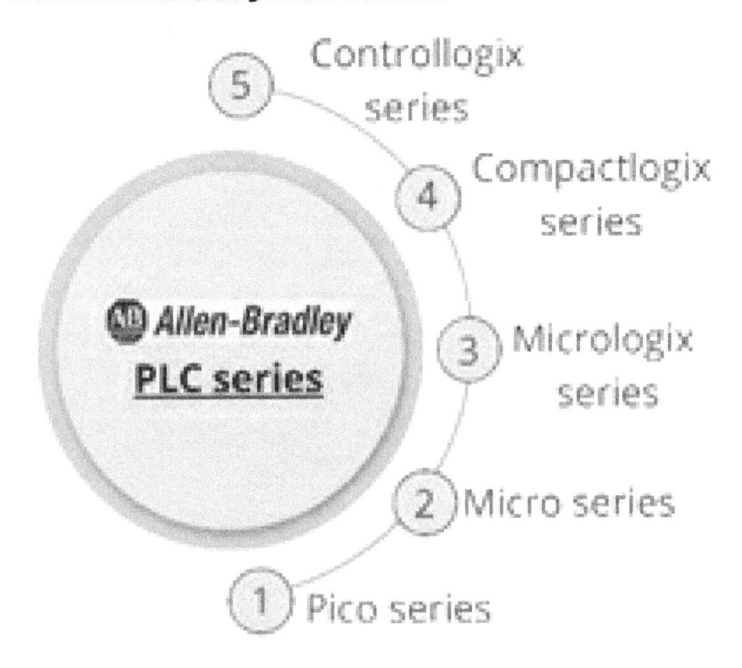

- 5. Controllogix series
- 4. Compactlogix series
- 3. Micrologix series
- 2. Micro series
- 1. Pico series

1. Pico PLC

Bulletin 1760 Pico Programmable Logic Controllers and Pico GFX Controllers are discontinued and no longer available for sale. AB recommends these replacement products by Micro Series PLC.

2. Micro800 control system

a. Micro810 Controllers

These controllers support 12 I/O points with 4 high current relay outputs (8A) for smart relay applications.

b. Micro820 Controllers

These controllers support up to 36 I/O points with many embedded features such as Ethernet, micro SD slot for recipe and data log, and analog I/O.

c. Micro830 Controllers

These controllers support up to 88 I/O points with high-performance I/O, interrupts, and PTO motion.

d. Micro850 Controllers

These controllers support up to 132 I/O points with high-performance I/O, interrupts, and PTO motion plus embedded Ethernet and 2085 expansion I/O.

3. MicroLogix control system

a. MicroLogix 1000 Controllers: Discontinued and no longer available for sale.

b. MicroLogix 1100 Controllers: The built-in LCD panel shows controller status, I/O status, and simple operator messages. With 2 analog inputs, 10 digital inputs and 6 digital versions. You can expand the I/O count using rack less I/O modules.

c. MicroLogix 1400 Controllers: Controllers without embedded analog I/O points provide 32 digital I/O points, while analog versions offer 32 digital I/O points and 6 analog I/O points. You can expand all versions with up to seven 1762 expansion I/O modules.

d. MicroLogix 1500 Controllers: Controller Systems will be discontinued and no longer available for sale outputs, the MicroLogix 1100 controller can handle a wide variety of tasks.

e. MicroLogix 1200 Controllers: The controller is available in 24-point and 40-point

4. ControlLogix control system

The ControlLogix family was introduced in 1997. This platform was racked-based having much faster scan times (speed) and memory than the PLC-5 or SLC products. Communication modules supported Ethernet, Device Net, DH485, and ControlNet.

1. ContrtolLogix 5570
2. ControlLogix 5580

5. CompactLogix control system

The CompactLogix family was released in 2008 as a lower-cost solution to ControlLogix for competitive reasons. And like the MicroLogix, the products do not use a rack-based solution but instead, use add-on modules to the ends of the power supply or CPU modules.

1. Compact guardLogix 5370 Safety controllers
Ideal for small, to mid-size applications that require low axis motion and I/O point counts
2. Compact GuardLogix 5380 Safety controllers
3. Compactlogix 5370 controllers
4. CompactLogix 5380 controllers
Ideal for small to midsize applications that require low axis motion and I/O point counts
5. CompactLogix 5480 controllers

B. Allen Bradley PLC programming software

There are three PLC programming software are used worldwide. But you need other software such as RS Linx classic to establish communication with the modules.

1. RS Logix 500
PLC- MicroLogix
Programming languages:
- Ladder logic

2. RS Logix 5000 (V20 to V30)

Programming languages:
- Ladder logic
- Structured Text
- Functional block
- Sequencial flow chart

3. Studio 5000 logix designer (V30 to V35)

Studio 5000 Logix designer is one of the best PLC programming software offer by Rockwell automation.
PLC- ControlLogix and CompactLogix

Salient features of studio 5000

- Safety instructions
- Motion control
- Third paty devices configuration

It is possible to configure any third party module with PLC with ease using EDS file or ethernet generic module.

Programming languages:

It supports various PLC programming languages
- Ladder logic
- Structured Text
- Functional block
- Sequencial flow chart

Controllers

1. Compact GuardLogix 5370 Safety controllers
2. Compact GuardLogix 5380 Safety controllers
3. Compactlogix 5370 controllers
4. CompactLogix 5380 controllers
5. CompactLogix 5480 controllers

6. ControlLogix 5570
7. ControlLogix 5580
8. GuardLogix 5570
9. GuardLogix 5580

NOTE: I've written a book specifically about the studio 5000 Logix designer. If that interests you as well, you can refer to it.

C. Communication protocols

Let's study different types of communication protocols suitable for Allen Bradley PLCs. These communication protocols can be used to establish a communication between PLC and PC or PLC and different devices.

1. DF1 protocol

It goes about the communication protocol designated for the classic serial link (RS232/RS422). The protocol allows the communication of either Full-duplex (point-to-point type of connection, RS232) or Half-duplex with addressing the PLC stations. This protocol is supported by devices that work with data areas (Data File).
Example: SLC 500, Micro Logix, etc.

2. Ethernet/IP protocol

It is built on the NetLinx architecture that is implemented even in DeviceNet and ControlNet networks. Ethernet/IP is administered by the ODVA independent organization that even takes charge of DeviceNet. This protocol is supported by devices that work with data objects (contrary to the data areas in DF1 protocol).
Example: Micro 820, control Logix, compact Logix, soft Logix, drive Logix, guard Logix, etc.

3. Modbus protocol

Modbus is a standard, very commonly used protocol for the

serial link and Ethernet (the PLC is slave). The Modbus RTU protocol is supported for example by: Micro820, MicroLogix 1200/1100/1500. The Modbus TCP/IP protocol is supported. Example: Micro 820, Micro Logix 1400.

4. ASCII protocol

It goes about the configurable char communication protocol designated for the classic serial link (RS232/RS422). This protocol is supported by some PLC's.
Example: Micro 820, compact Logix (on the second serial link port), micro Logix 1100/1200/1500, etc.

5. DH+ protocol

It supports 64 nodes at the most, 230 kB/s (SLC 501 to 3 can't it). It is the proprietary Allen- Bradley protocol. Two methods can be used for communication.

6. DH485 protocol

It goes about the communication protocol designated for the serial link. It supports 32 nodes at the most, 19 kB/s (multi-master = each- to-each). It is the proprietary Allen-Bradley protocol. Two methods can be used for communication.

Matching IP addresses

The IP in Ethernet/IP stands for Internet protocol. This communication method is typically implemented via ethernet port located on the device you use, but if the devices have wireless capabilities it can also be done through wireless communication. An IP address follows the structure of xxx.xxx.xxx.xxx with each set of xxx equating a number between 0 to 255.

In order to communicate between your computer and PLC, you will have to ensure that your computer is on the same network as the device. Changing your computer's IP address is fairly simple and can be done by following steps:

- In the computer's control panel, click on the Network and

Internet Button

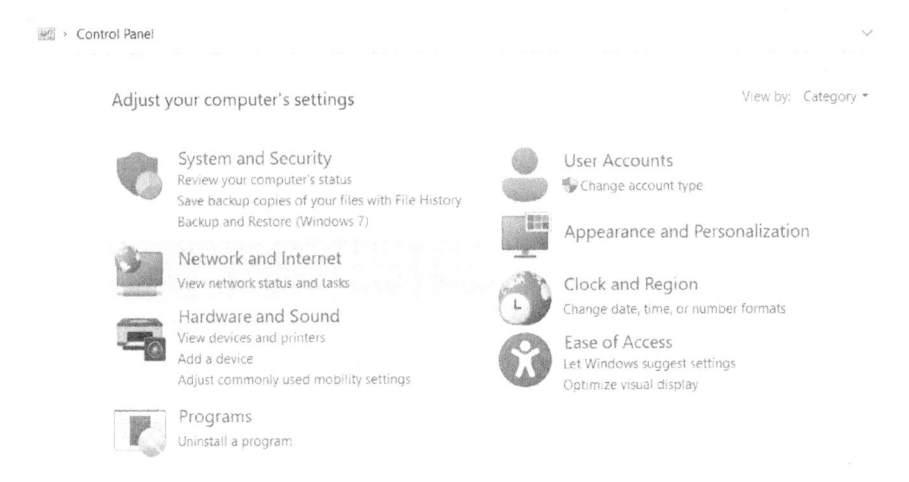

- Then, navigate to the Network and sharing center.

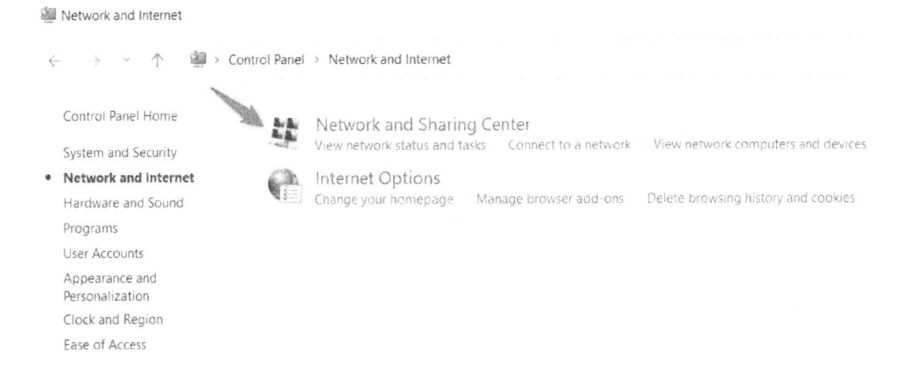

- Once in the Network and sharing center, click on the change adapter settings option.

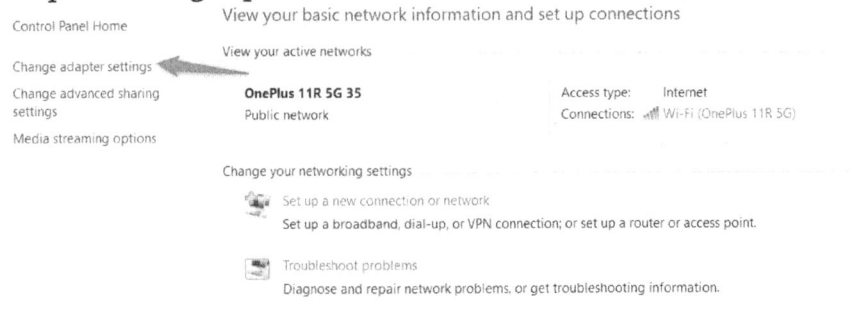

- From here, navigate to the Ethernet properties menu by double clicking on the Ethernet button.

- In this menu, click on the Internet Protocol Version 4 (TCP/IPv4) list them and click on properties.

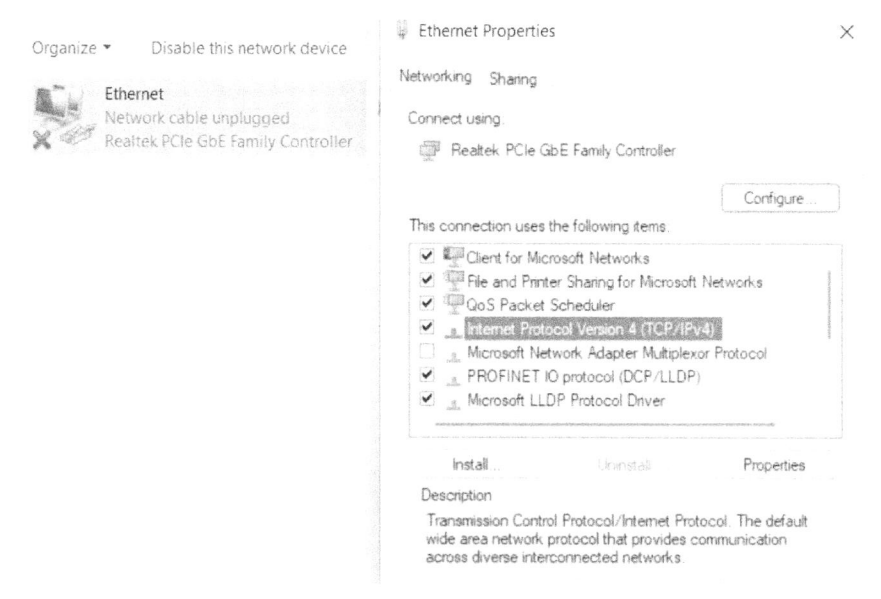

- In this menu, you can set a static IP address for your computer.

The first three sets of IP addresses you enter when configuring your computer's IP address should correspond to the IP address of the device you want to connect to. If, for instance, your PLC's IP address is 192.168.10.10, then your computer's IP address needs to be changed to something like 192.168.10.12. You can ping the IP address of the PLC from your computer after setting the IP address.

D. RS Linx classic

RSLinx® Classic

Whenever you use any programming software from Allen Bradly family, you have to use RS Linx software to communicate with PLC and other devices. RS linx classic is an industrial communication software for Rockwell automation networks and devices. Without this software you cannot even download a PLC program into the PLC.

Configure a driver

Cables for communication between PLC and computer.
- Ethernet cable
- USB cable

Click on communication and then select configure drivers. A window will appear as shown above in front of you, you have to select a driver that you are using. If you don't have PLC training kit, then you can select emulator to check your written program.

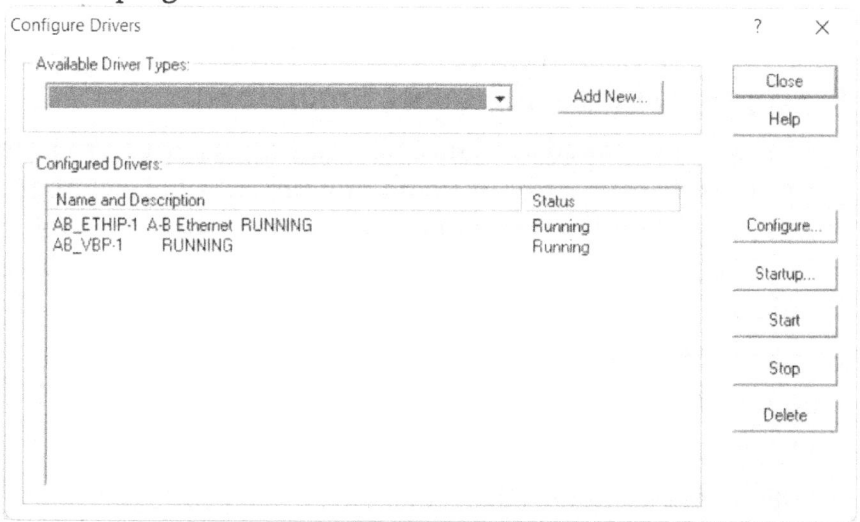

After selecting a driver, you can see its status. The status of driver must be in running condition.

Who active

After configuration of drivers, you can view a list of all the connected devices with their IP addresses. Not only ethernet devices but also USB devices can be seen in RS Linx classic software.

To check the connected devices, go to communication and select 'who active'.

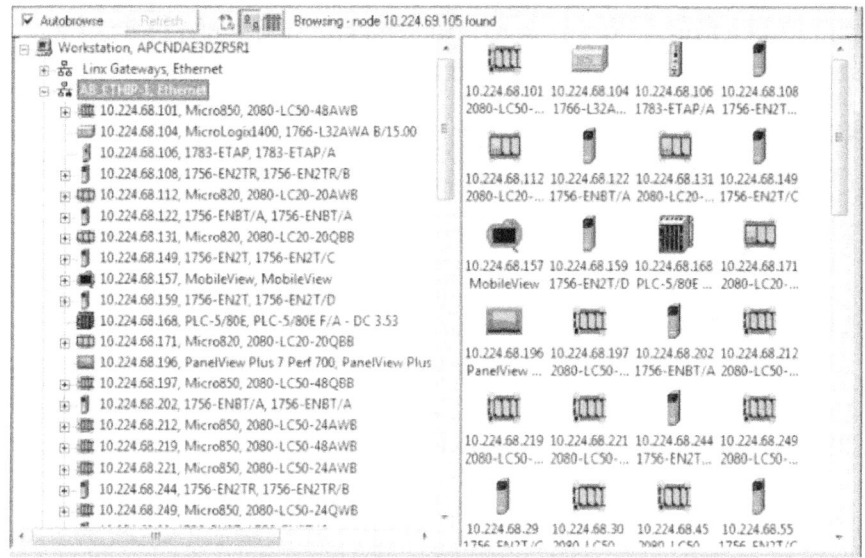

E. Program simulation using Emulator

A virtual PLC called an emulator can be used to emulate your PLC software without using any physical hardware. An emulator is the ideal instrument for testing your reasoning prior to deployment. In this section, we will learn how to simulate a program using the studio 5000 Emulator.

We need software:

1. Studio 5000 Emulator

2. RS Linx traditional V4.0

3. 5000 Logix designer V30

4. Studio Factory TalkView V12

1. In Studio 5000 Emulate, create a virtual PLC. Launch the Studio 5000 emulator. You can choose any of the 16 available controller slots, with the exception of slot 0.

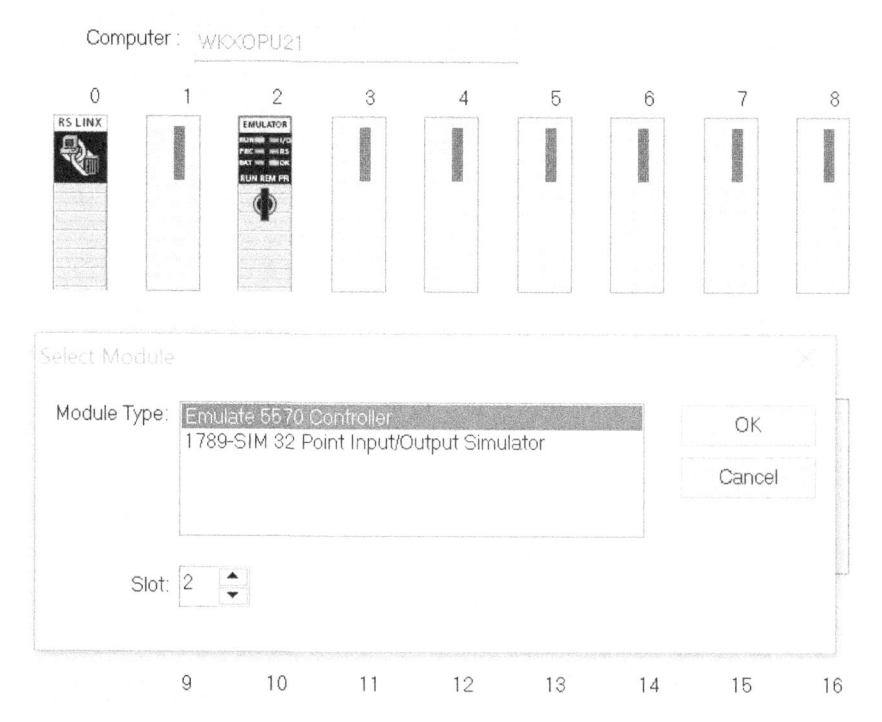

Right-click any open slot, then select the Create tab.
A new window will appear, select Emulate 5570 controller along with slot number.

General

Type: Studio 5000® Logix Emulate EmuLogix5868
Vendor: Allen-Bradley
Version: 30

Startup Mode: Remote Program

Memory Size (KB): 3072

Periodic Save Interval: 10
(Range: 0.5 to 30 min)   ☐ Enable Periodic Save

Controller Name: Last Loaded:

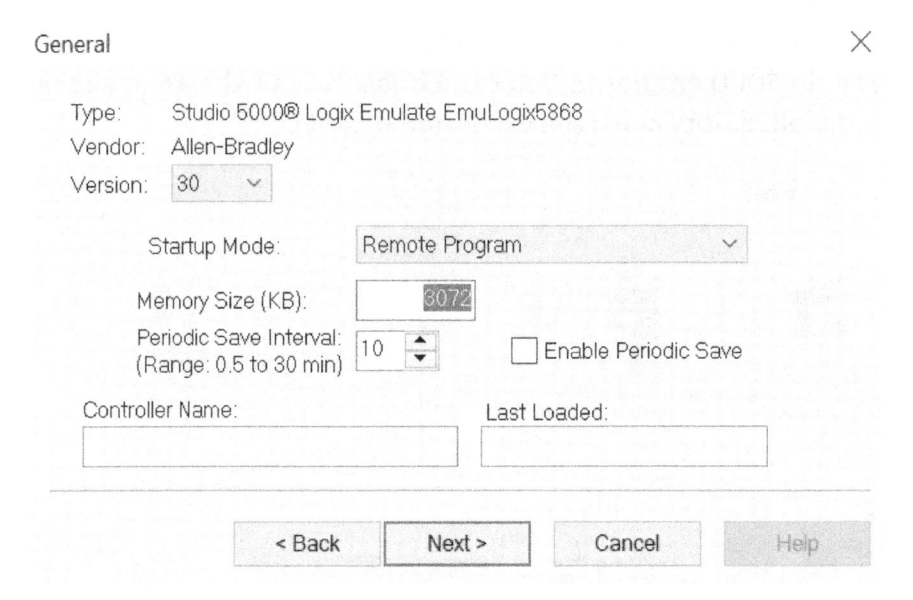

< Back   Next >   Cancel   Help

2. Configure a driver for virtual PLC

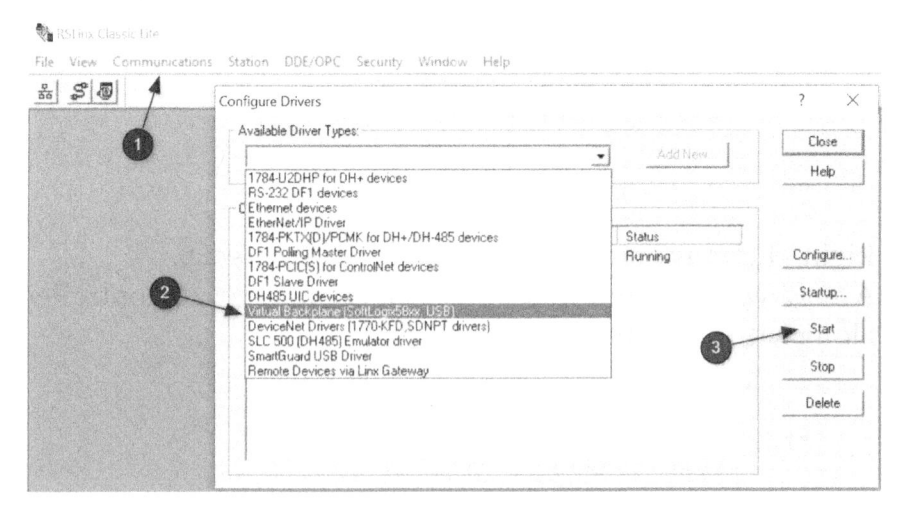

1. In the communication tab, click on configure driver
2. Select Virtual Backplane
3. Click on start tab

3. Download a program and go online

Controller setting

In order to download a program into the virtual PLC, you must change the controller type in studio 5000 Logix designer.

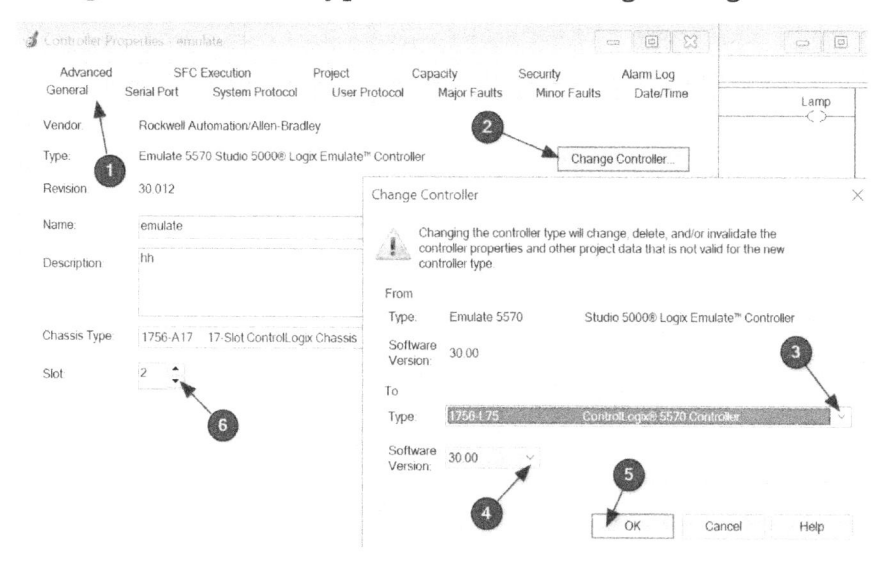

1. Right click on plc and select properties. Click on general tab.
2. Click on change controller
3. Select Emulator controller
4. Keep software of studio 5000 logix designer that will go with studio 5000 emulate software.
5. Click OK
6. Select the slot where the virtual PLC is placed.

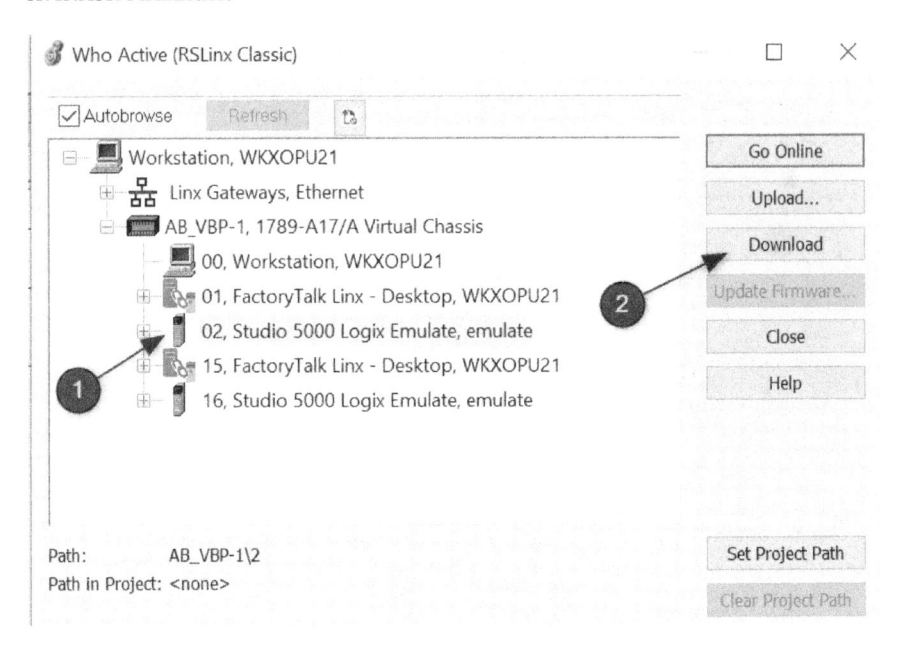

1. In the communication tab, select who active and select the virtual PLC.

2. Click on download.

GO Online

1. Here you can see the communication path.

2. Click on Rem run tab and select Run.

3. The program is in Run mode now, you can see the power lines in green color.

4. Configure Virtual PLC with Factory TalkView

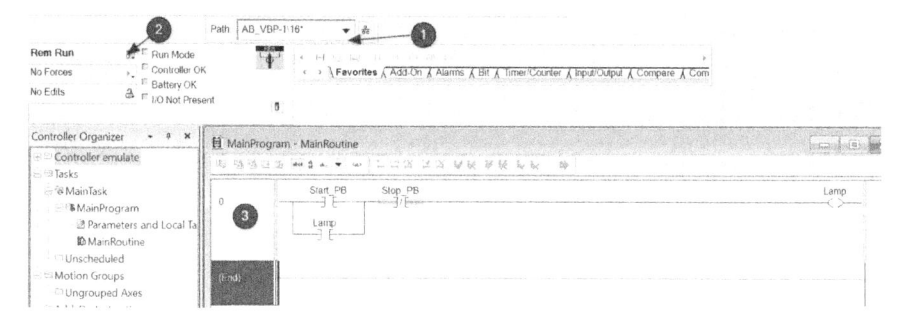

Communication setup

1. Click on communication setup
2. Select the Virtual PLC and click on OK

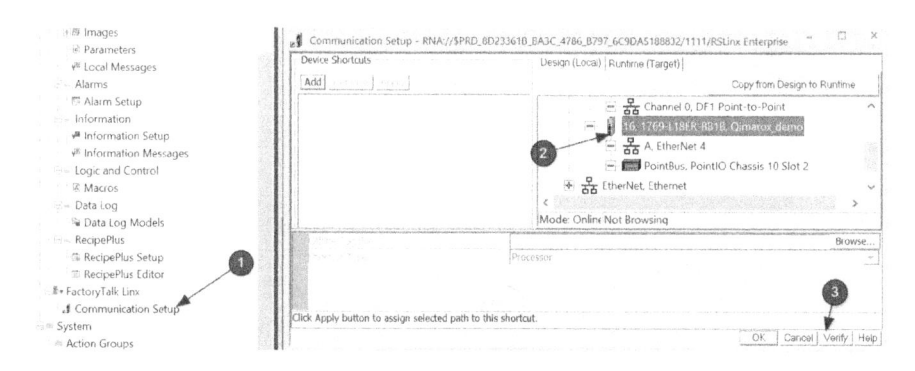

Assign tags

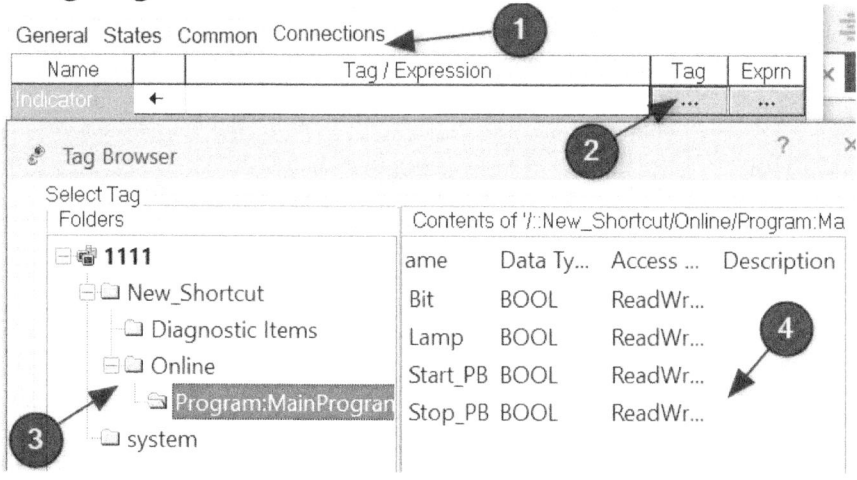

37

1. Right click on the Push button and select communication.
2. Click on the three dots.
3. If PLC is online, you can see all the tags in the left part after clicking on online folder.
4. Select a tag.

Simulation

HMI and PLC software can both be simulated simultaneously. You may watch the entire program as it runs. From the simulation, you can perform the necessary actions.

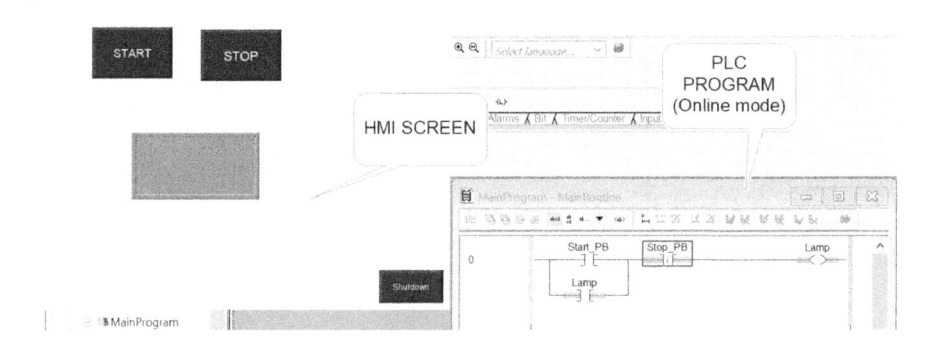

2.1 RS LOGIX 500

A. Introduction

RSLogix 500

Rockwell Automation

The RSLogix 500 software is a tool to design and implement ladder programs for the Allen-Bradley SLC 500 and MicroLogix family of processors. It is a Windows based application produced by Rockwell Software, that allows PLC programmation using a personal computer.

Several other features of RSLogix 500, such as a project verifier, drag-and-drop editing, and search-and-replace functions, greatly facilitate PLC programming. The PLC can be programmed via either a RS-232 port or an Ethernet port on the PLC processor.

This software comes with RSLinx which provides connectivity between the PLC and the PC-type computer. The software is available with either an educational license (Model 3245-C) or as a commercial version (Model 3245-D).

Overview screen
1. Programming window
This window is for writing a ladder logic. It has vertical power line within which you have to develop ladder logic.

2. Tool Bar

The tool bar has a treasure of ladder logic instructions that you just need to drag and drop.

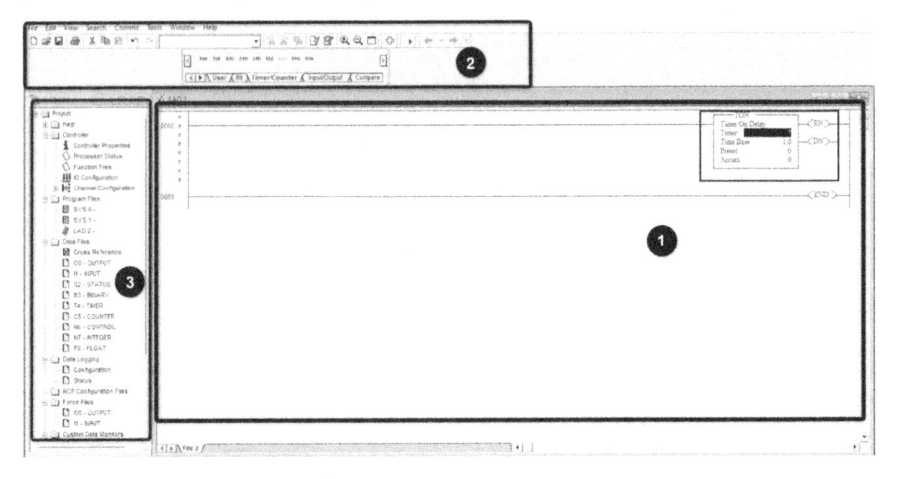

3. Navigation bar

a. Data Files

In an RSLogix 500 application, the user will find currently specified data structures in the IO tree under the section called "Data Files" (See below). Note that this is different from RSLogix 5000.

You will notice that every data structure is specified by a letter, an integer as well as string.

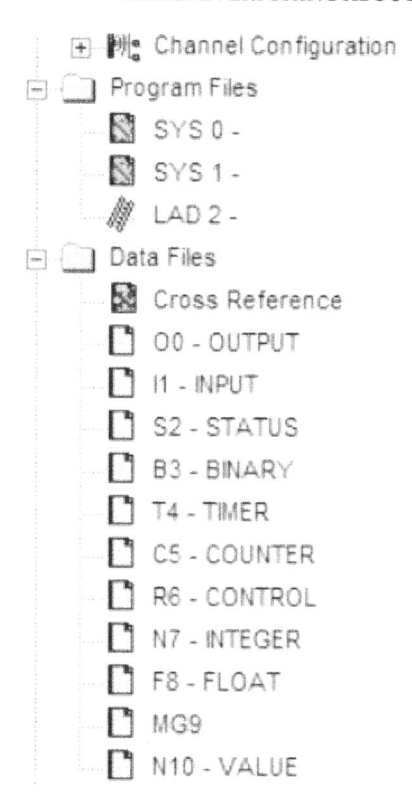

```
⊞ ▦ Channel Configuration
⊟ ▢ Program Files
    ▨ SYS 0 -
    ▨ SYS 1 -
    ▦ LAD 2 -
⊟ ▢ Data Files
    ▨ Cross Reference
    ▢ O0 - OUTPUT
    ▢ I1 - INPUT
    ▢ S2 - STATUS
    ▢ B3 - BINARY
    ▢ T4 - TIMER
    ▢ C5 - COUNTER
    ▢ R6 - CONTROL
    ▢ N7 - INTEGER
    ▢ F8 - FLOAT
    ▢ MG9
    ▢ N10 - VALUE
```

Types of data

O0 Output- This file stores the state of output terminals for the controller.

I1 Input- This file stores the state of input terminals for the controller.

S2 Status- This file stores controller operation information useful for troubleshooting controller and program operation.

B3 Bit- This file stores internal coil's status.

T4 Timer- This file stores the timer accumulator and preset values and status bits.

C5 Counter- This file stores the counter accumulator and preset values and status bits.

N7 Integer- This file is used to store bit information or numeric values with a range of -32767 to 32768.

F8 Floating-point This file stores a # with a range of

1.1754944e-38 to 3.40282347e+38.

Understanding Specific Data Types

Now that we've covered the different data types, it's time to give more information with regards to their layout.

When the user right-clicks a data structure and selects "Properties", he will be presented with the same configuration window as the one when a new data structure is created. In this window, the user will find multiple key parameters.

B. Getting started with RS Logix 500

1. Select a PLC

Whenever you open the RS Logix 500 software, it asks you to select the series of PLC from the list.

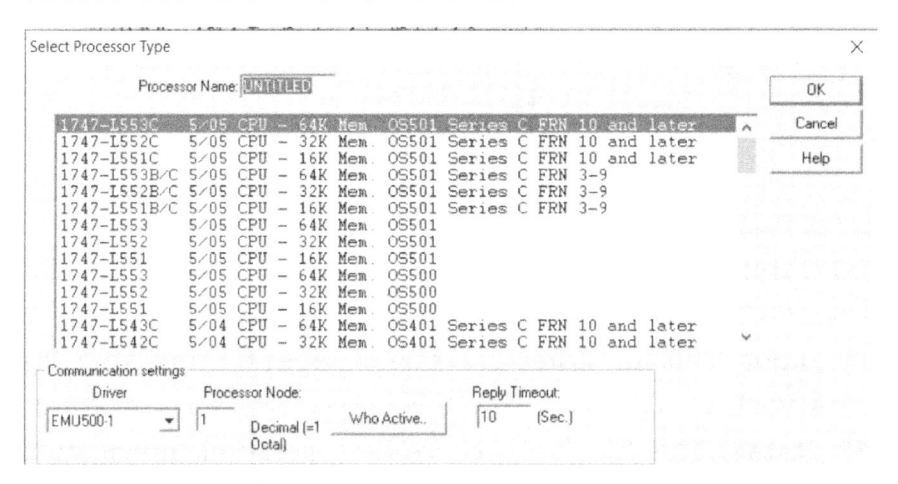

2. Add a module

As per the different application's need you have to add different modules with the existing PLC. You can configure different modules such as digital, analog, etc with your PLC. The modules can be added in two ways: Manually and Automatically.

LEARN EVERYTHING ABOUT PLC PROGRAMMING

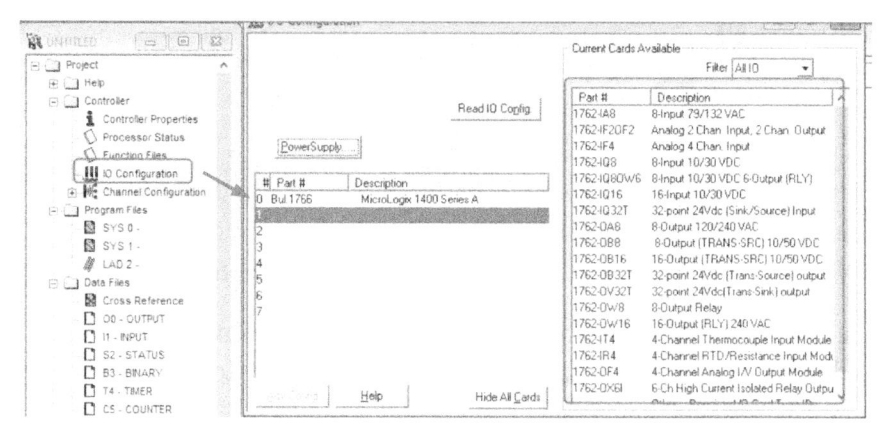

As you can see a Micro Logix 1400 PLC has bee assigned to slot 0.

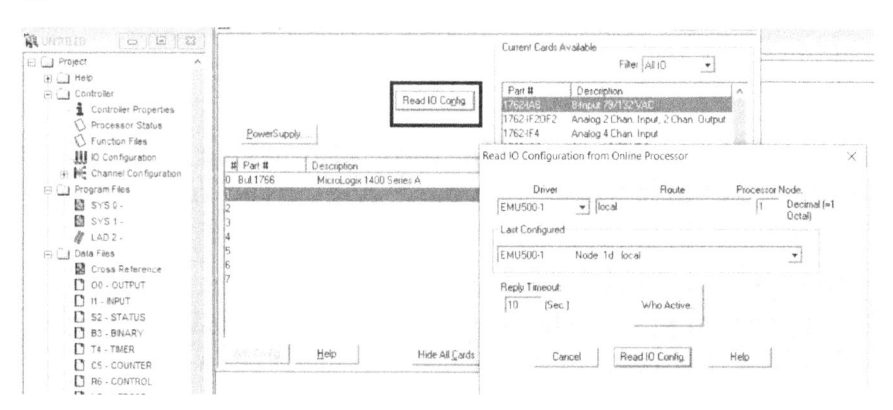

- When you select modules, the slots get assigned to them.
- The modules can be configured automatically by clicking on Read IO confi.

3. Program writing procedure

Creating the 1st Rung of Ladder Logic

a) Add an Input Instruction

43

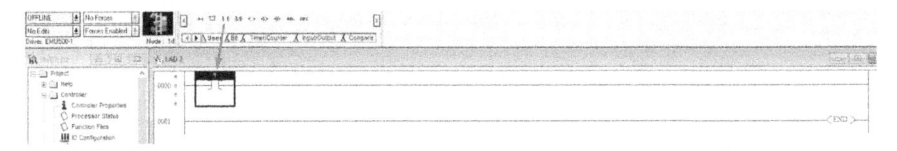

Click, hold the left mouse button and drag the XIC (examine if closed) button onto the left side of the rung you just created. When you see a green box, release the mouse button.

With the instruction highlighted Type I:0/0 [Enter]. This is the address of the XIC (examine if closed) instruction.

b) Add Output Instruction

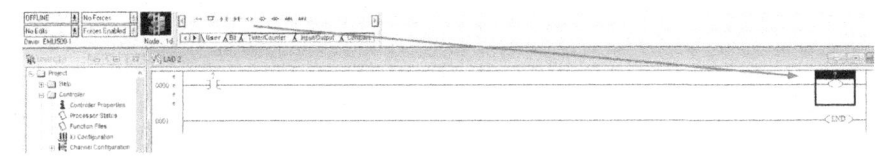

Click, hold the left mouse button and drag the OTE (output energized) button onto the right side of the rung you just created.

When you see a green box, release the mouse button. With the instruction highlighted Type O:0/0 [Enter]. This is the address of the OTE instruction.

Creating the 2nd Rung of Logic

a) Create a New Rung

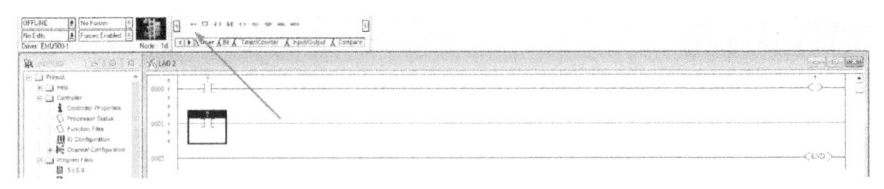

Click, hold the left mouse button and drag the "New Rung" button over rung "0001".

When you see a green box, release the mouse button.

b) Add the 1st Input Instruction

Click, hold the left mouse button and drag the XIO (examine if open) button onto the left side of the rung you just created.

When you see a green box, release the mouse button.

4. Verify your work

If Program has errors

To find the errors in the program click on the error message in the "Verify results window" the error is then highlighted in the ladder window.

Fix the error and run "Verify file" again.

When all the errors are fixed you can then save and download the program.

5. Documenting your Program

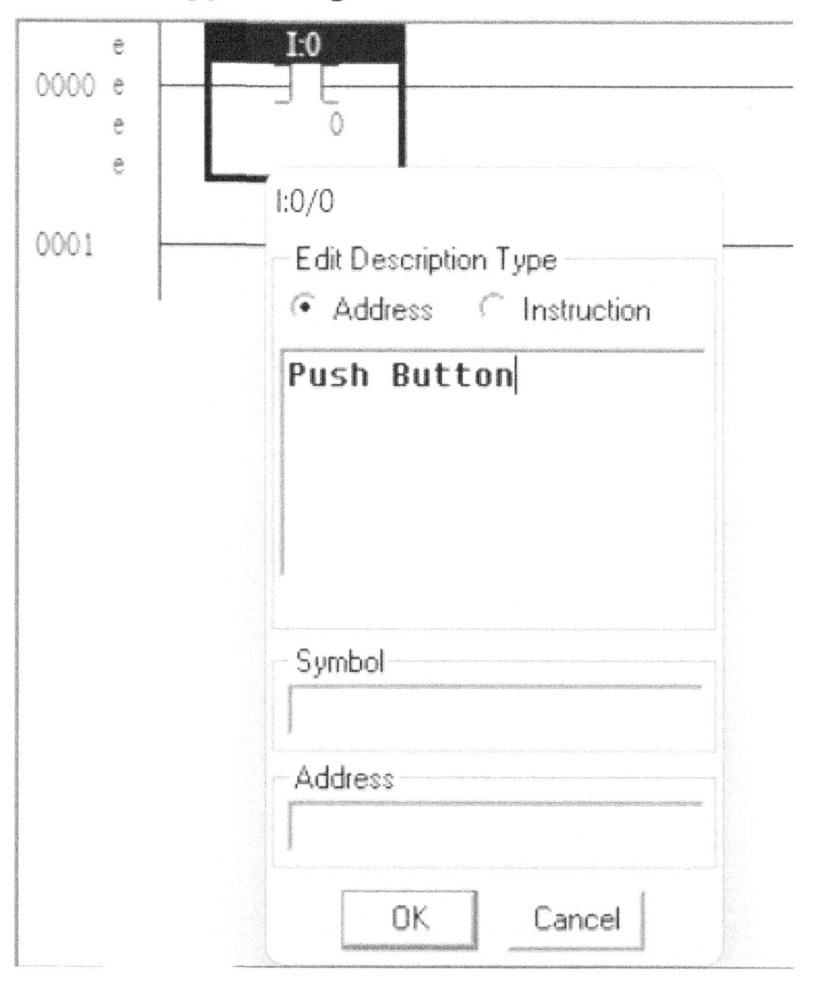

1. Click on the Input I:0/0 to highlight
2. Right mouse on I:0/0 and select "Edit Description- I:0/0"
3. Select "Address"
4. Type "Input Switch 0" in the Edit window and Select "OK".

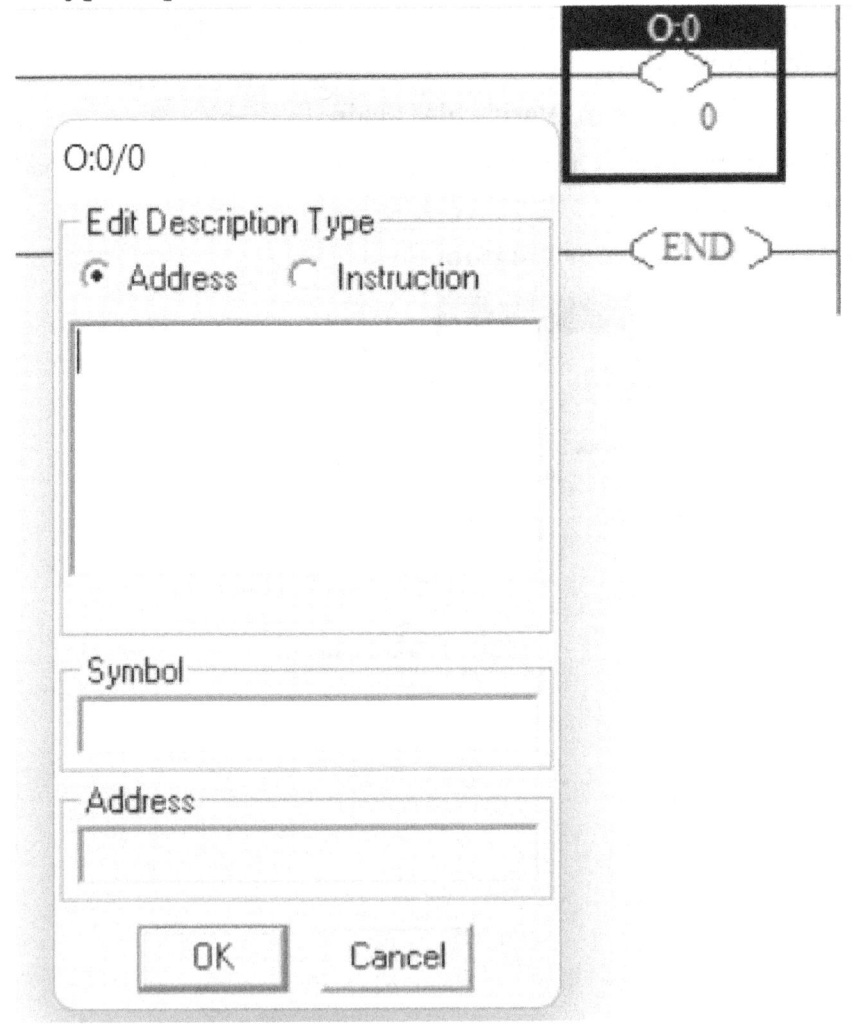

5. Click on the Output O:0/0 to highlight
6. Right mouse on O:0/0 and select "Edit Description- O:0/0"
7. Select "Address"
8. Type "Output 0" in the Edit window

9. Select "OK"

Download the Program

- Select the menu item "Comms > System Comms"
- Three primary selections
- "Online" Establish the "path"
- "Upload" Receive from the controller
- "Download" Send to the controller
- Highlight the device at Node 01.
- Select "Download

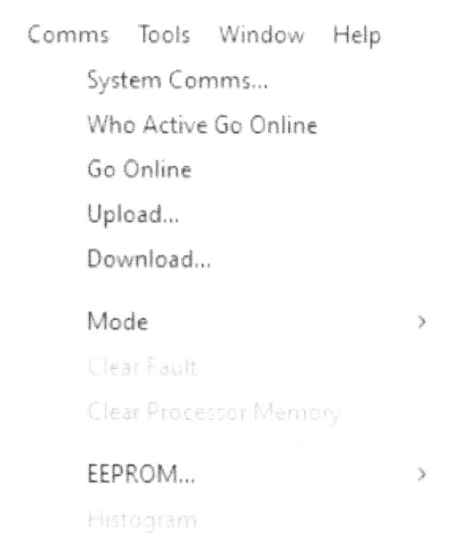

```
Comms   Tools   Window   Help
        System Comms...
        Who Active Go Online
        Go Online
        Upload...
        Download...
        Mode                    >
        Clear Fault
        Clear Processor Memory
        EEPROM...               >
        Histogram
```

Select "OK" in the Revision note window. The revision note box is used to keep track of changes made to the existing program. You can create many revisions of the same program. This feature can be disabled if desired.

Select "Yes" to download your program over the existing program that resides in the processor. This window will appear whenever a program is being downloaded to the processor. The download verification window will appear as downloading completes.

Put the Micro into the RUN mode

Select "Yes" to go online. This will allow you to monitor the program that now resides in the processor.

1. Click on the down arrow next to the word REMOTE PROG
2. Three Run selections
3. One scan cycle with outputs disabled
4. Select "Run"
5. Select "Yes" for "Are you sure window

You just have to right click on the bit you wish to troubleshoot, a cross reference window will get opened. You can easily trace that OTE to solve the problem.

C. Ladder logix instructions

In this section we are going to see different ladder logic instructions that we need for programming. These instructions are same in RS Logix 500 and studio 5000 Logix designer.

1. Bit instructions

1. XIO
- Examines a bit for an off condition.
- Use an XIO instruction in your ladder logic to determine if a bit if off.
1 = True
0 = False

Devices
Start/Stop push buttons, Selectors, Limit switch, Proximity switch, Light, Internal bit, etc.

2. XIC
- Examines a bit for an on condition
Use the XIC instruction in your ladder logic to determine if a bit is ON.
0 = False
1 = True

Devices

Start/Stop push buttons, Selectors, Limit switch, Proximity switch, Light, Internal bit, etc.

3. OTE

- Turns a bit on or off.
- Use OTE instruction in your ladder logic to turn on a bit when rung condition is evaluated as true.

Devices
Light, Motor run signal, Internal coil, etc.

Milk bottle filling and capping application

Start the conveyor in auto mode. When a bottle is present at sensor 1, the conveyor stops and the controller turns on the valve after 3 seconds, and starts filling the bottle up to a certain level (Untill level sensor is true), then the valve gets off and the conveyor starts again.

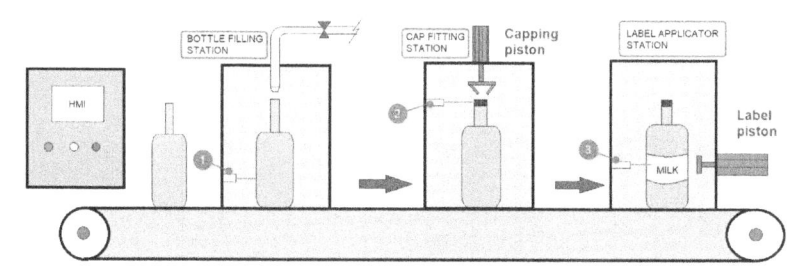

When a bottle reaches at capping station, the conveyor gets stopped, then the capping piston operates, once the cap is detected by a sensor, the capping piston stops and the conveyor starts again.

When a bottle reaches at labelling sensor, the conveyor stops, and the labeling piston operates, when a label is detected by a sensor, the labeling piston stops and the conveyor starts again. This cycle continues until a stop PB is pressed.

Input devices
- Start PB
- Stop PB

- Auto/Man switch
- Bottle present sensor 1
- Level sensor
- Bottle present sensor 2
- Cap present sensor
- Bottle present sensor 3
- Label present sensor

Output devices

- Conveyor motor
- Solenoid valve
- Capping piston
- Label piston

Written ladder logic

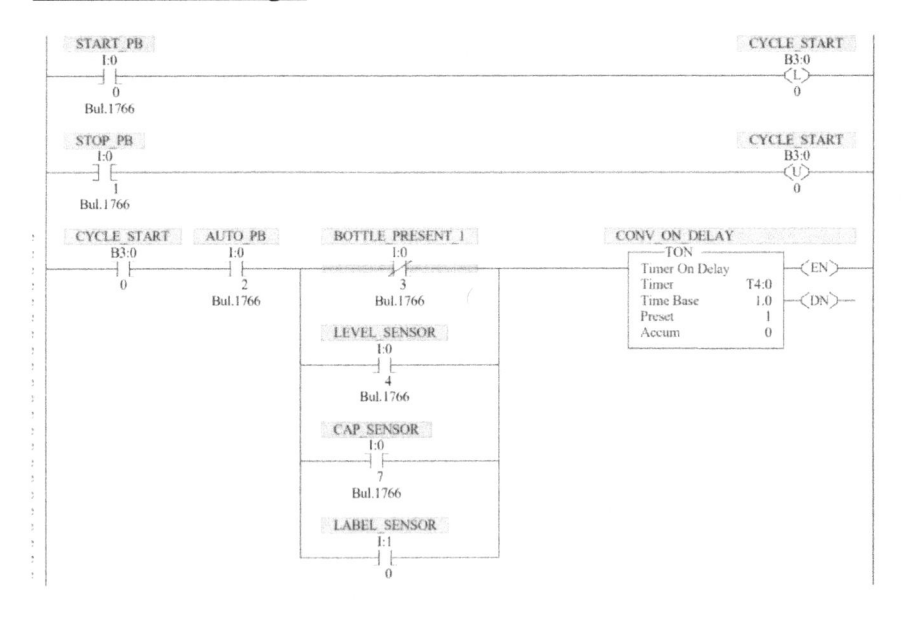

LEARN EVERYTHING ABOUT PLC PROGRAMMING

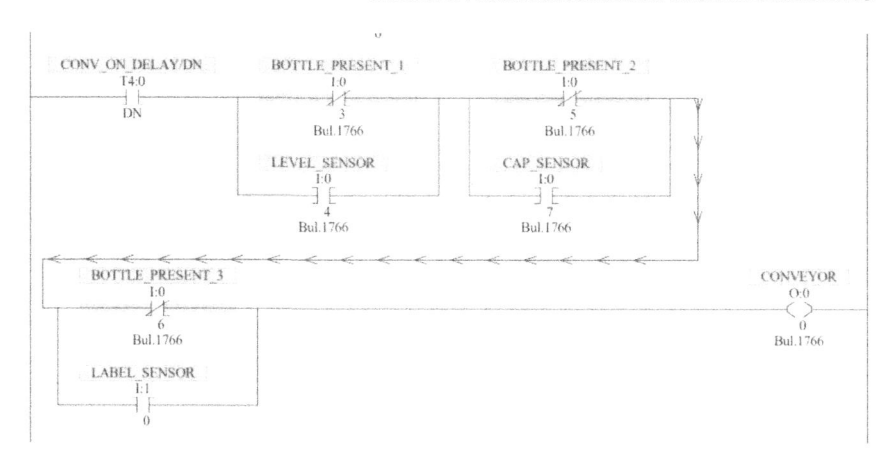

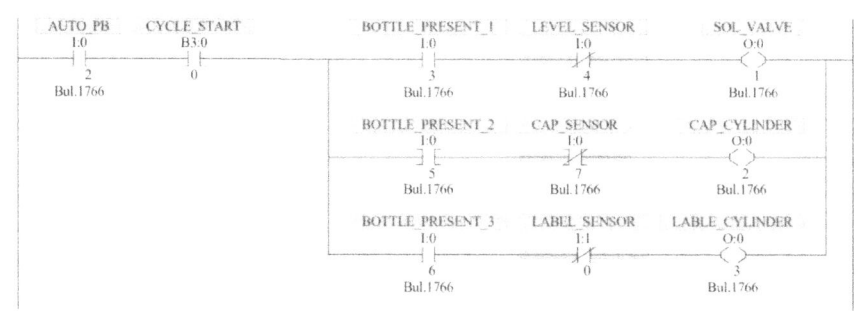

```
   CONV_ON_DELAY/DN      BOTTLE_PRESENT_1     BOTTLE_PRESENT_2
        T4:0                    I:0                  I:0
   ─┤ ├─────────────────┬──┤/├──────────────┬──┤/├──────────────
         DN             │    3              │    5
                        │  Bul.1766         │  Bul.1766
                        │  LEVEL_SENSOR     │  CAP_SENSOR
                        │       I:0         │       I:0
                        └──┤ ├──────────────┴──┤ ├──
                               4                    7
                            Bul.1766            Bul.1766

   BOTTLE_PRESENT_3                                            CONVEYOR
        I:0                                                      O:0
   ─┬──┤/├───────────────────────────────────────────────────────( )─
    │    6                                                        0
    │  Bul.1766                                                 Bul.1766
    │  LABEL_SENSOR
    │      I:1
    └──┤ ├──
           0


   AUTO_PB   CYCLE_START    BOTTLE_PRESENT_1   LEVEL_SENSOR      SOL_VALVE
     I:0        B3:0              I:0              I:0              O:0
   ─┤ ├────┤ ├──────────┬──┤ ├──────────────┬──┤/├──────────────────( )─
     2       0          │    3              │    4                    1
   Bul.1766             │  Bul.1766         │  Bul.1766             Bul.1766
                        │  BOTTLE_PRESENT_2 │  CAP_SENSOR           CAP_CYLINDER
                        │       I:0         │       I:0                O:0
                        ├──┤ ├──────────────┼──┤/├──────────────────( )─
                        │    5              │    7                    2
                        │  Bul.1766         │  Bul.1766             Bul.1766
                        │  BOTTLE_PRESENT_3 │  LABEL_SENSOR         LABLE_CYLINDER
                        │       I:0         │       I:1                O:0
                        └──┤ ├──────────────┴──┤/├──────────────────( )─
                               6                    0                    3
                            Bul.1766            Bul.1766             Bul.1766
```

2. Latch / Unlatch instructions

1. OTL

This instruction functions much the same as the OTE with the exception that once a bit is set with an OTL, it is "Latched" on, once an OTL bit has been set "ON" (1 in Memory) it will remain "ON" even if the rung condition goes false the bit be reset with an OTU instruction. Latch and unlatch instructions must be assigned the same address in your logic program.

2. OTU

Use this output instruction to unlatch (Reset) a latched (Set) bit which was set by an OTL instruction. The OTU address is identical to the OTL address which originally set the bit.

4. OSR and OSF instructions

- The state of this contact is true when a positive transition

(OFF to ON) is detected on assigned bit I0.0.

- The bit M0.0 will be ON for 1 clock cycle.

```
                POSITIVE                         NEGATIVE
                 EDGE                              EDGE
ON      ─────────┐ ┌──────────────────────────────┐ ┌─────
                 │ │                              │ │
OFF              └─┘                              └─┘
                M0.0                             M0.1

ON      ────────┌──────────┐───────────┌──────────┐──────
                │          │           │          │
OFF     ────────┘          └───────────┘          └──────
               I0.0                   I0.1              TIME
```

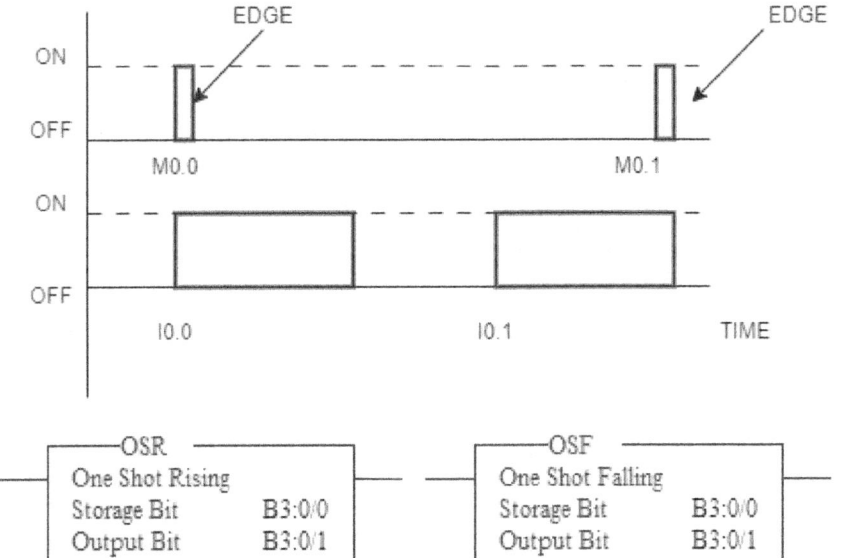

─OSR─		─OSF─	
One Shot Rising		One Shot Falling	
Storage Bit	B3:0/0	Storage Bit	B3:0/0
Output Bit	B3:0/1	Output Bit	B3:0/1

- The state of this contact is true when a negative transition (ON to OFF) is detected on assigned bit I0.0.

- The bit M0.0 will be ON for 1 clock cycle. These bits can be connected anywhere except output.

3. Timer

A timer is used to delay in the process or complete the process in a certain timer period. Timers are available readily into the programming software, we just need to use it with time value. There are three types of timer used in RS Logix 500 software.

a. On delay timer

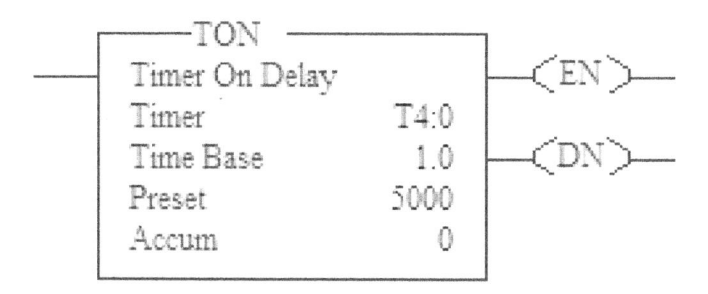

```
         ┌─── TON ──────────────┐
         │  Timer On Delay      │──( EN )──
     ────│  Timer         T4:0  │
         │  Time Base      1.0  │──( DN )──
         │  Preset        5000  │
         │  Accum            0  │
         └──────────────────────┘
```

Use the TON instruction to turn an output on or off after the timer has been on for a preset time interval. This output instruction begins timing (at either one second or one hundredth of a second intervals) when its rung goes "true." It waits the specified amount of time (as set in the PRESET), keeps track of the accumulated intervals which have occurred (ACCUM), and sets the DN (done) bit when the ACCUM (accumulated) time equals the PRESET time.

b. Off delay timer

Use the TOF instruction to turn an output on or off after its rung has been off for a preset time interval. The TOF instruction begins to count timebase intervals when the rung makes a true-to-false transition.

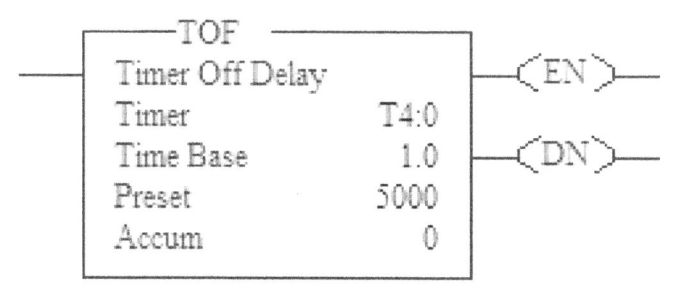

```
         ┌─── TOF ──────────────┐
         │  Timer Off Delay     │──( EN )──
     ────│  Timer         T4:0  │
         │  Time Base      1.0  │──( DN )──
         │  Preset        5000  │
         │  Accum            0  │
         └──────────────────────┘
```

As long as rung conditions remain false, the timer increments its accumulated value (ACC) based on the timebase for each scan until it reaches the preset value (PRE). The Accumulated value is reset when rung conditions go true regardless of whether the timer has timed out.

c. Retentive timer

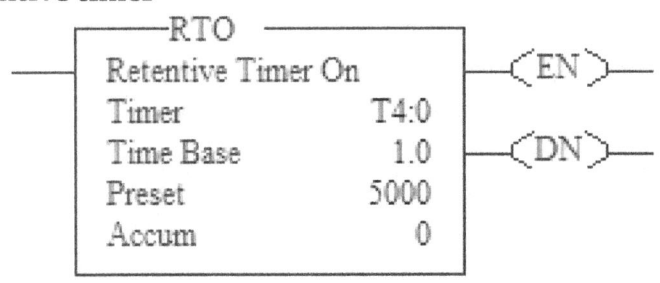

```
      ─RTO─────────────
──────┤ Retentive Timer On     ├──(EN)──
      │ Timer          T4:0    │
      │ Time Base       1.0    ├──(DN)──
      │ Preset         5000    │
      │ Accum             0    │
      └────────────────────────┘
```

An RTO function the same as a TON with the exception that once it has begun timing, it holds its count of time even if the rung goes false, a fault occurs, the mode changes from REM Run or REM Test to REM Program, or power is lost. When rung continuity returns (rung goes true again), the RTO begins timing from the accumulated time which was held when rung continuity was lost. By retaining its accumulated value, retentive timers measure the cumulative period during which rung conditions are true.

4. Counter

A counter is used to count the bits. When a single bit is available, a counter counts it as one and keeps counting the bits. There are two types of timer used in RS Logix 500 software.

a. Count up (CTU)

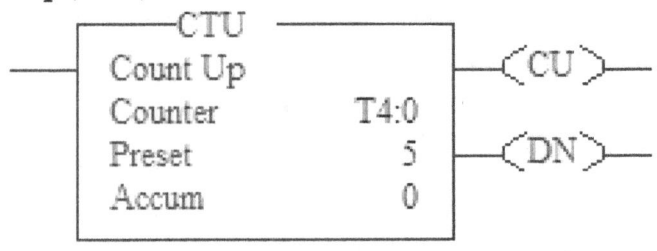

```
      ─CTU─────────────
──────┤ Count Up              ├──(CU)──
      │ Counter        T4:0   │
      │ Preset            5   ├──(DN)──
      │ Accum             0   │
      └───────────────────────┘
```

This output instruction counts up for each false-to-true transition of conditions preceding it in the rung and produces an output when the accumulated value reaches the preset value. Rung transitions might be triggered by a limit switch or by parts traveling past a detector.

The ability of the counter to detect false-to-true transitions depends on the speed (frequency) of the incoming signal. The on and off duration of an incoming signal must not be faster than the scan time.

Each count is retained when the rung conditions again become false, permitting counting to continue beyond the preset value. This way you can base an output on the preset but continue counting to keep track of inventory/parts, etc.

Note: Use a RES (reset) instruction with the same address as the counter, or another instruction in your program to overwrite the value. The on or off status of counter done, overflow, and underflow bits is retentive. The accumulated value and control bits are reset when a RES is enabled.

Enter a COUNTER address, PRESET value and ACCUM value. The preset value is the point which must be reached to set the DN (done) bit. The accumulated value represents the current count status. C5:1 represents counter file number 5, element number 1.

b. Count down (CTD)

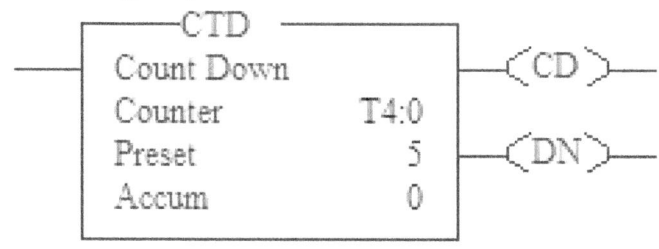

```
        ┌──CTD─────────────────┐
────────┤ Count Down           ├──( CD )──
        │ Counter         T4:0 │
        │ Preset             5 ├──( DN )──
        │ Accum              0 │
        └──────────────────────┘
```

This output instruction counts down for each false-to-true transition of conditions preceding it in the rung and produces an output when the accumulated value reaches the preset value. Rung transitions might be triggered by a limit switch or by parts traveling past a detector.

Each count is retained when the rung conditions again become false. The count is retained until a RES (reset) instruction with the same address as the counter is enabled, or if another instruction in your program overwrites the value.

The accumulated value is retained after the CTU or CTD instruction goes false, and when power is removed from and then restored to the processor. Also, the on or off status of counter done, overflow, and underflow bits is retentive. The accumulated value and control bits are reset when a RES is enabled.

5. Math instructions

You need mathematical calculation during programming, RS Logix 500 has basic math instructions like Add, Sub, Mul, Div, Square root, Modulo, and compute. Each of these math instructions uses two variables.

The data types you can use in math instructions are SINT, INT, DINT, Floats, and REALs.

a. Add instruction

Use the ADD instruction to add one value to another value (Source A + Source B) and place the sum in the destination.

b. Subtraction instruction

Use the SUB instruction to subtract one value from another value (Source A - Source B) and place the result in the destination.

c. Multiply instruction

Use the MUL instruction to multiply one value with another value (Source A * B) and place the result in the destination.

d. Divide instruction

Use the DIV instruction to divide one value by another value (Source A / B) and place the result in the destination.

6. Compare instructions

The compare instructions are test instructions that compare two numbers or strings. At least one of the values in each comparison must be a tag value. For example comparison between one integer N1 with another integer N2, comparison between integer value N1 and real value 20.0, etc. The compare instructions compare source A with source B and then produces output.

Here are the popular compare instructions we use in ladder logic.

- Equal
- Not equal
- Limit
- Greater than
- Less than
- LIM instruction

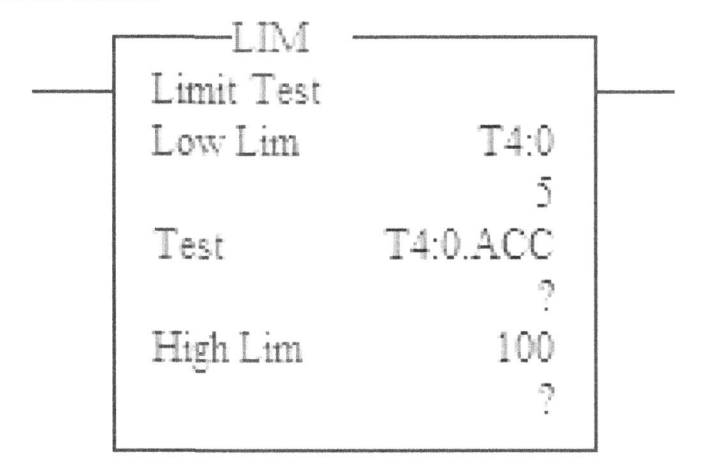

```
┌─ LIM ─────────────────────┐
│  Limit Test               │
│  Low Lim          T4:0    │
│                      5    │
│  Test          T4:0.ACC   │
│                      ?    │
│  High Lim          100    │
│                      ?    │
└───────────────────────────┘
```

The limit instruction gets true when the test value is is within the specified limits.

7. Move and copy instructions

a. Move instruction

```
┌──── MOV ─────────────┐
│    Move              │
│    Source      N4:0  │
│                   5  │
│    Dest     T4:0.PRE │
│                   ?  │
└──────────────────────┘
```

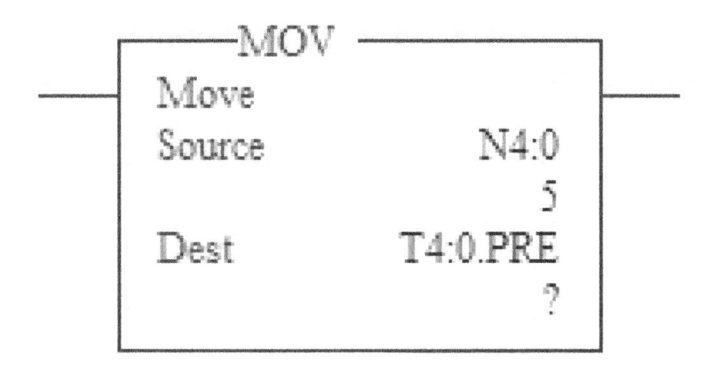

Source and Destination can be different data sizes. The source is converted to the destination size when the instruction executes. If the signed value of the Source does not fit in the destination, the overflow is handled as follows:

1. If the Math Overflow Selection Bit is clear, a saturated result is stored in the Destination. If the Source is positive, the Destination is 32767 (word). If the result is negative, the Destination is -32768.

2. If the Math Overflow Selection Bit is set, the unsigned truncated value of the Source is stored in the Destination. The source can be a constant or an address.

b. Copy instruction

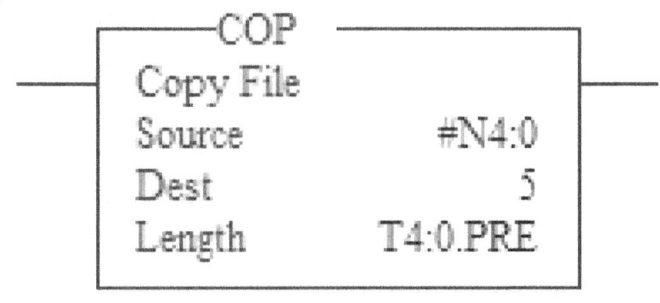

```
┌──── COP ─────────────┐
│    Copy File         │
│    Source     #N4:0  │
│    Dest            5 │
│    Length   T4:0.PRE │
└──────────────────────┘
```

The COP instruction copies blocks of data from one location into another. The source and destination file types must be the same except bit (B) and integer (N); they can be interchanged.

It is the address that determines the maximum length of the block to be copied.

3. Clear instruction

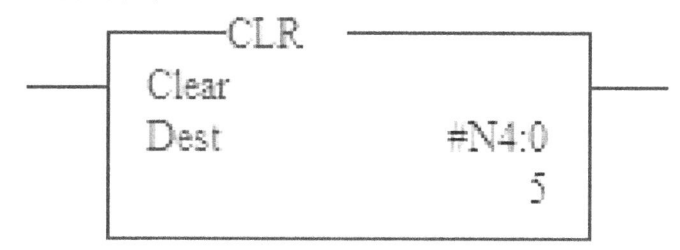

```
         ─CLR─────────────────
        │ Clear               │
     ───│ Dest         #N4:0  │───
        │                  5  │
         ─────────────────────
```

The clear instruction is used to clear the integer value.

8. Logical Instruction

You may be familiar with logic gates which you might have studied before. The studio 5000 logix designer has different logic instructions as shown below.

a. AND

This instruction gets enabled when both source A and source B bits are available.

b. OR

This instruction gets enabled when one of the sources or both sources are available.

c. XOR

This instruction works exactly opposite to OR gate.

d. NOT

This instruction gets enabled when the source bit is false.

9. PID instruction

The analog sensors are used to recieve accurate data from the system, in that case the signal varries from 0 to 20 dc volts or 4 to 20 mA. b

This analog signal needs to be converted into the data word.

To convert the analog data, the following formula is used.

N=Vin x 1023/10

Where, Vin (analog signal) is in volts (V)

Analog signal	Data word
0V	0
5V	512
10V	1023

<u>Scale with parameter</u>

The SCP instruction produces a scaled output value that has a linear relationship between the input and scaled values. This instruction solves the following equation listed below to determine scaled output. The analog sensor is assigned with analog input is I0:1.1.

$$y = [(y1-y0) / (x1-x0)] (x-x0) + y0$$

```
      ┌──SCP──────────────┐
──────┤ Scale w/Parameters├──────
      │ Input        I:0.0│
      │                  ?│
      │ Input Min.       0│
      │                  ?│
      │ Input Max.   17500│
      │                  ?│
      │ Scaled Min.      0│
      │                  ?│
      │ Scaled Max.  17500│
      │                  ?│
      │ Output       O:0.0│
      │                  ?│
      └───────────────────┘
```

Boiler application

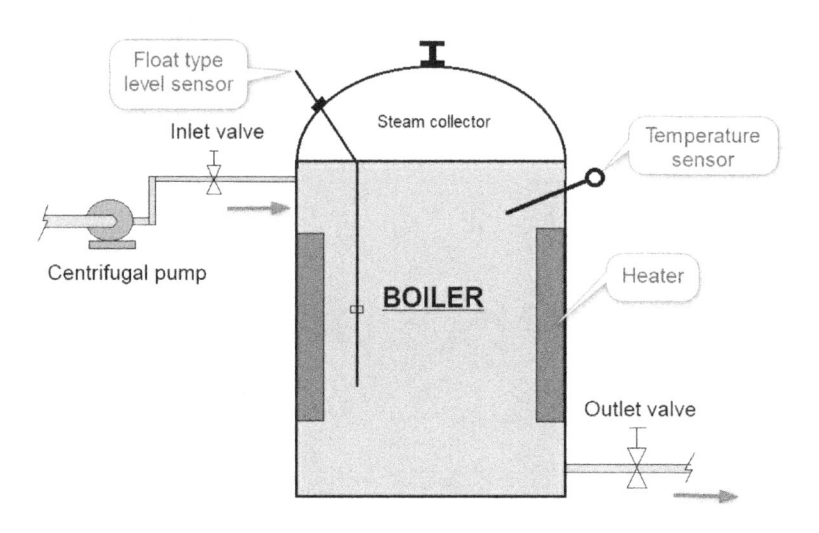

The boiler process starts when a start button is pressed from SCADA. If the level of water in the boiler is less than 20 ltr, the inlet valve will get on first and the centrifugal pump will get on after 3 sec. And they remain on until the water level reaches 80 ltr.

Once the water level reaches 80 ltr, the pump and inlet valve will stop. The heater will start heating the water inside the boiler.

As soon as the temperature reaches 110 degrees Celsius, the heater will turn off. The outlet valve will be opened from SCADA to drain the boiled water. The outlet valve remains open until the water level reaches 20 ltr. This cycle continues until a stop button is pressed.

<u>Input devices</u>
- Temperature sensor
- Level sensor

<u>Output devices</u>

LEARN EVERYTHING ABOUT PLC PROGRAMMING

- Inlet valve
- Outlet valve
- Centrifugal pump

Programming using RS Logix 500

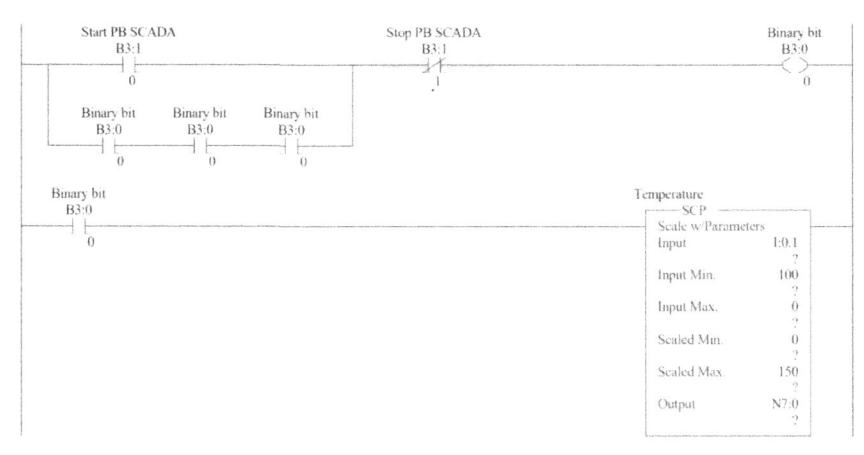

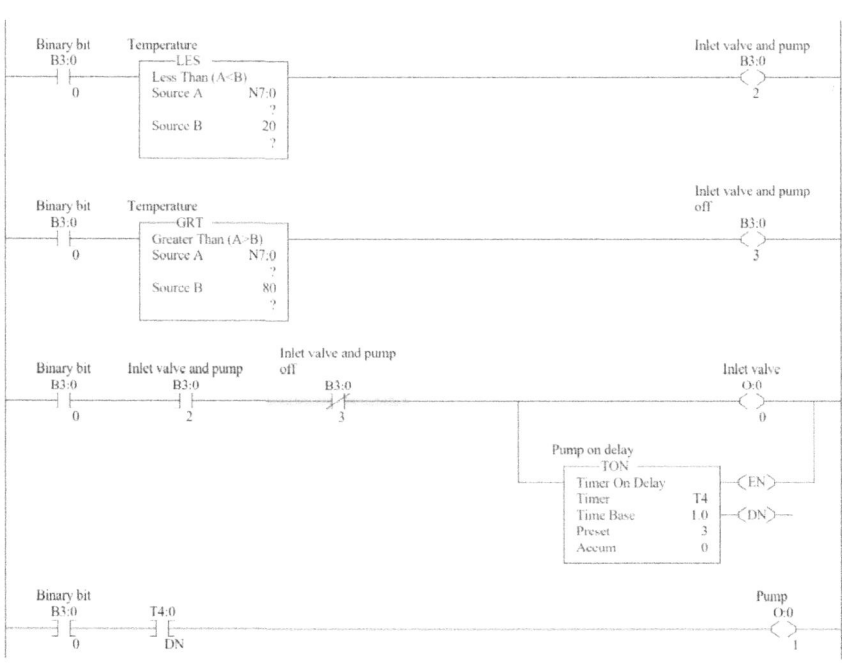

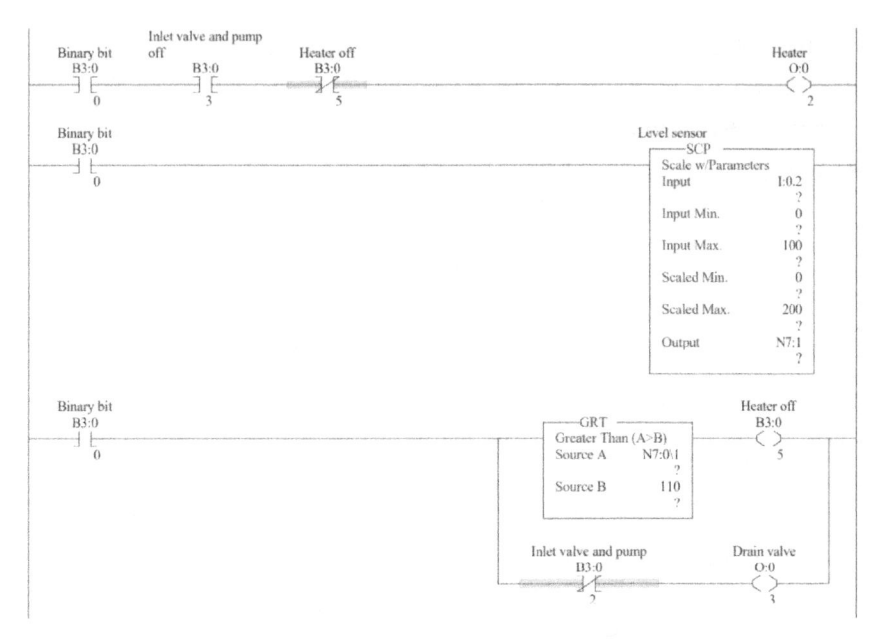

```
                Inlet valve and pump
  Binary bit    off              Heater off                                            Heater
    B3:0              B3:0          B3:0                                                O:0
────┤ ├───────────────┤ ├───────────┤/├──────────────────────────────────────────────────( )────
       0                 3             5                                                 2

  Binary bit                                                        Level sensor
    B3:0                                                     ┌──────────SCP──────────┐
────┤ ├─────────────────────────────────────────────────────┤  Scale w/Parameters    ├───
       0                                                    │  Input            I:0.2│
                                                            │                    ?   │
                                                            │  Input Min.        0   │
                                                            │                    ?   │
                                                            │  Input Max.       100  │
                                                            │                    ?   │
                                                            │  Scaled Min.       0   │
                                                            │                    ?   │
                                                            │  Scaled Max.      200  │
                                                            │                    ?   │
                                                            │  Output           N7:1 │
                                                            └────────────────────?───┘

  Binary bit                                                                          Heater off
    B3:0                                  ┌────────GRT────────┐                         B3:0
────┤ ├───────────────────────────────────┤ Greater Than (A>B)├─────────────────────────( )────
       0                                  │ Source A   N7:0\1 │                          5
                                          │              ?    │
                                          │ Source B    110   │
                                          │              ?    │
                                          └───────────────────┘

                                          Inlet valve and pump                        Drain valve
                                                  B3:0                                   O:0
                                          ────────┤/├────────────────────────────────────( )────
                                                   ?                                      3
```

10. Program control instructions

1. JSR instruction

The JSR instruction causes the controller to start executing a separate subroutine file within a ladder program. JSR moves program execution to the designated subroutine (SBR file number). After executing the SBR, control proceeds to the instruction following the JSR instruction.

2. Return instruction

The RET instruction marks the end of subroutine execution or the end of the subroutine file. It causes the controller to resume execution at the instruction following the JSR instruction, user interrupt, or user fault routine that caused this subroutine to execute.

11. Shift instructions

a. BSL (Bit Shift Left)

The BSL and BSR instructions are used to shift a bit to particular position. A BSL shifts a bit to left position, whereas BSR shifts a bit to the right position within the array.

The source bit's value is loaded into the array's lowest bit by this instruction and the unload bit is determined by the length.

D. Troubleshooting

With the controller in "Remote Run", you can monitor or edit data within the controller. Change data variables while in run When "Green" bars are shown on either side of logic elements, this indicates "Logical Continuity", this helps to determine how the application is operating. This design is to help in debugging an application's logic.

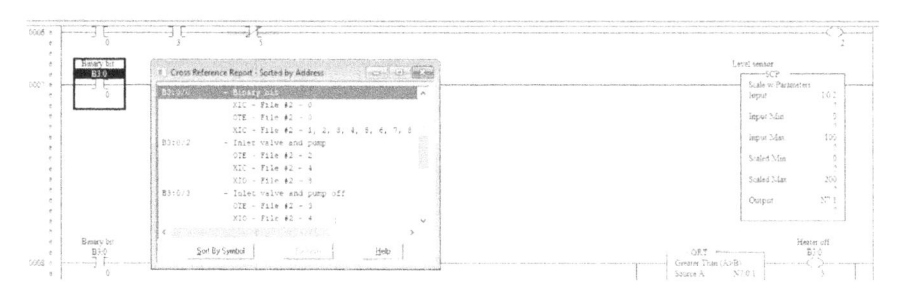

Online troubleshooting is easier than tracing a single wire separately. Let's consider your machine gets broken down during operation and you observe its pusher cylinder is not working. You can troubleshoot it by monitoring a program.

E. PanelView 800 HMI

CONTROL COMMANDS

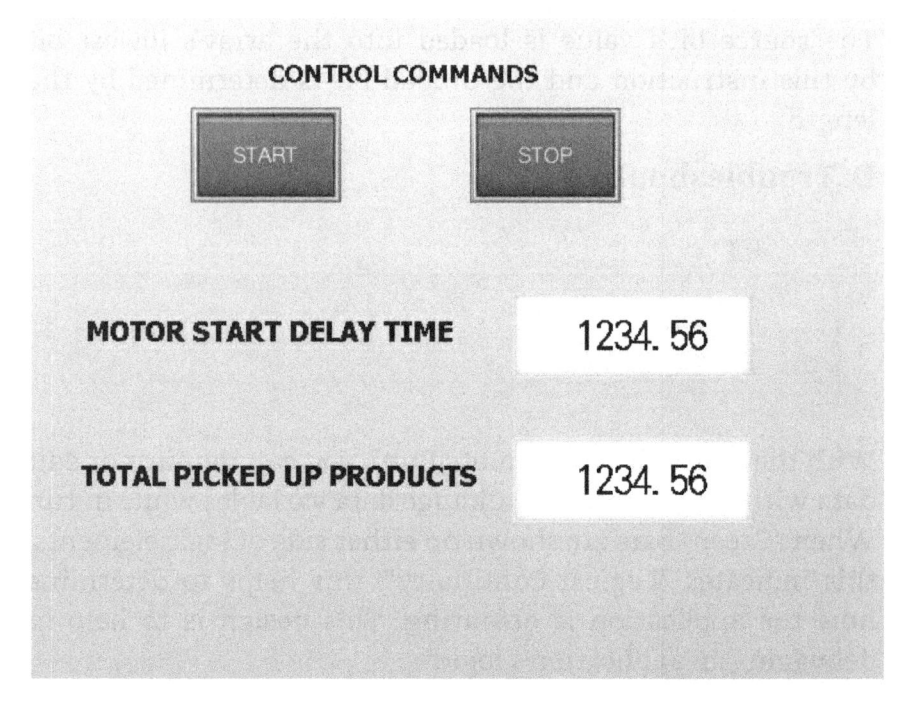

[START] [STOP]

MOTOR START DELAY TIME   1234.56

TOTAL PICKED UP PRODUCTS   1234.56

MicroLogix 1100 serial communication with panelview HMI using DF1. I have developed this screen using FT view software.

1. Control commands from HMI

```
Start PB    Stop PB                                    Conveyor
B3:0        B3:0                                         O:0
 ┤├─────────┤├─────────────────────────────────────────( )
  0           1                                          0
                                                       Bul.1766
```

A machine can be easily control through HMI by creating control tabs. As you can see above, two tabs have been created and two binary bits B3:0/0 and B3:0/1 have been assigned to start and stop push buttons.

2. Numeric entry from HMI

HMI provides with a facility to change the values of parameters such as Frequency, time, etc. To accept the value from outside we need to use numeric input tab. As the above rung shows, the time of motor start delay can be changed by entering value from HMI.

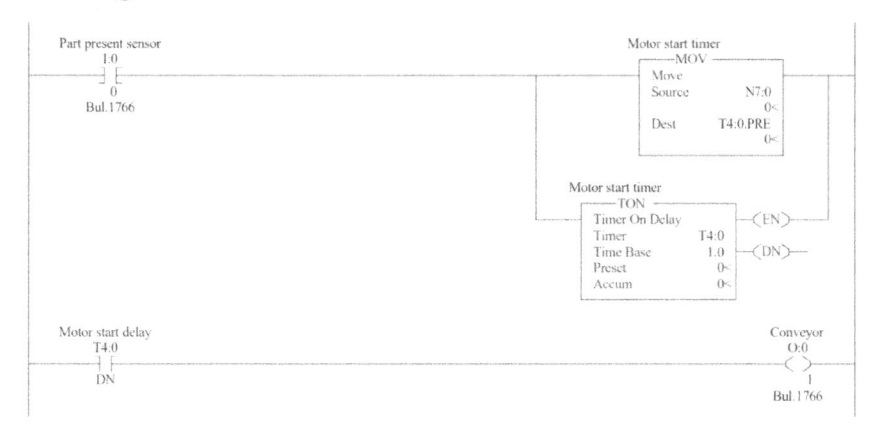

4. Display

When I0:0/0 bit gets true, counter keeps counting in an incremental manner. This incremental value is moved to N7:1 register which is assigned to the display tab on HMI. HMI displays the value as it keeps changing.

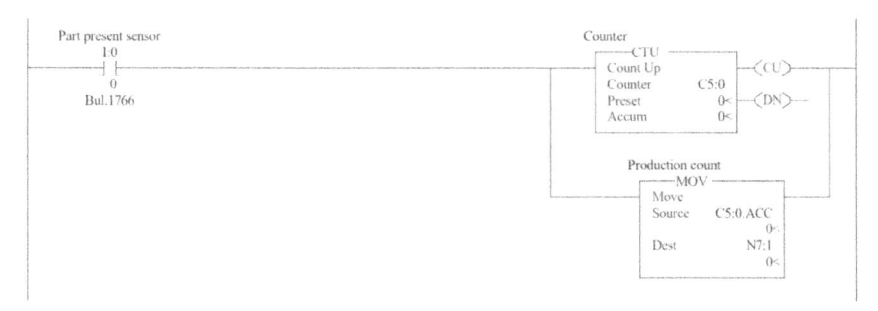

5. Alarm

As per the above rung, if I0:00 is present the on-delay timer T4:0/0 starts counting 20 seconds. If the product is not picked up within 20 seconds, then Alarm O0:0/0 should on.

F. RS Logix 500 and Powerflex 525 VFD

In order to control the power Flex 525 VFD with Modbus communication, we need to set a few communication parameters as shown in the list.

MicroLogix 1400 Modbus RTU Settings

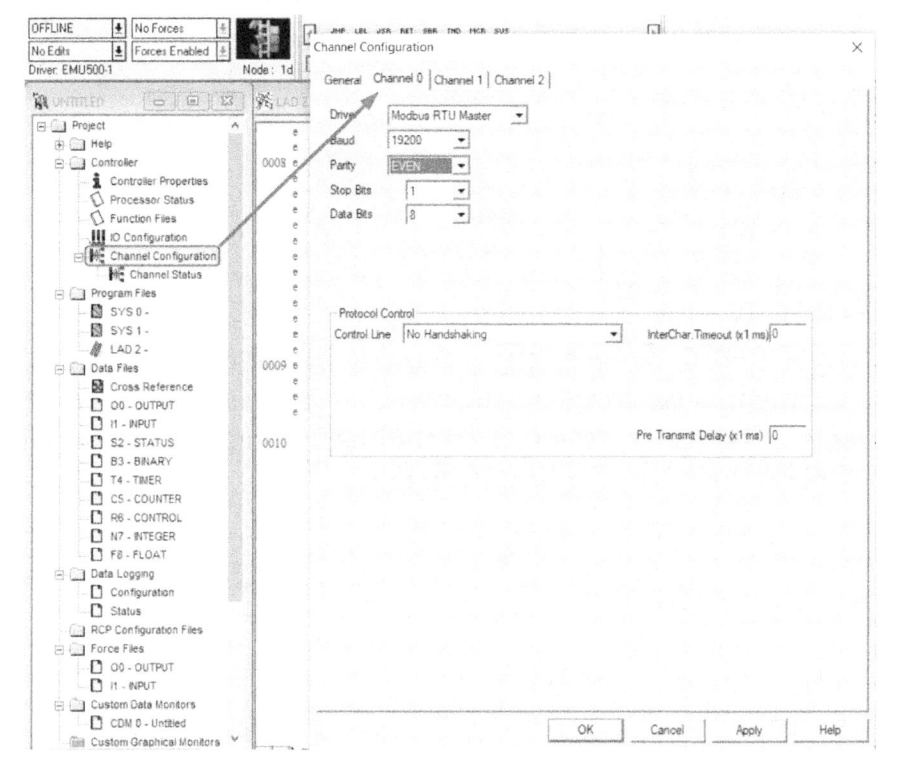

Parameters

The following parameters you need to set into VFD for the Modbus communication.

Parameter	Description	Set value	Remark
C121	Comm write mode	0	"save" when programming drive
C122	Com logic command	0	Velocity
C123	RS 485 Data Rate	3	9600 bps
C124	RS 485 Node Address	1	Slave 1
C125	Comm loss action	0	Fault
C125	Comm loss time	5	5 Seconds
C126	RS 485 format	1	Modbus RTU
P046	Start source 1	3	Start is controlled from the network
P047	Speed reference 1	3	Speed reference is controlled from the network

PanelView HMI Screen

Control commands- The VFD operations like forward, reverse, and stop can be controlled from HMI.

Speed control- The VFD speed can be controlled by entering frequency from HMI.

Data monitoring- Data like frequency, speed, current can be shown on HMI screen.

RS Logix 500 programming

Read / Write Distribution

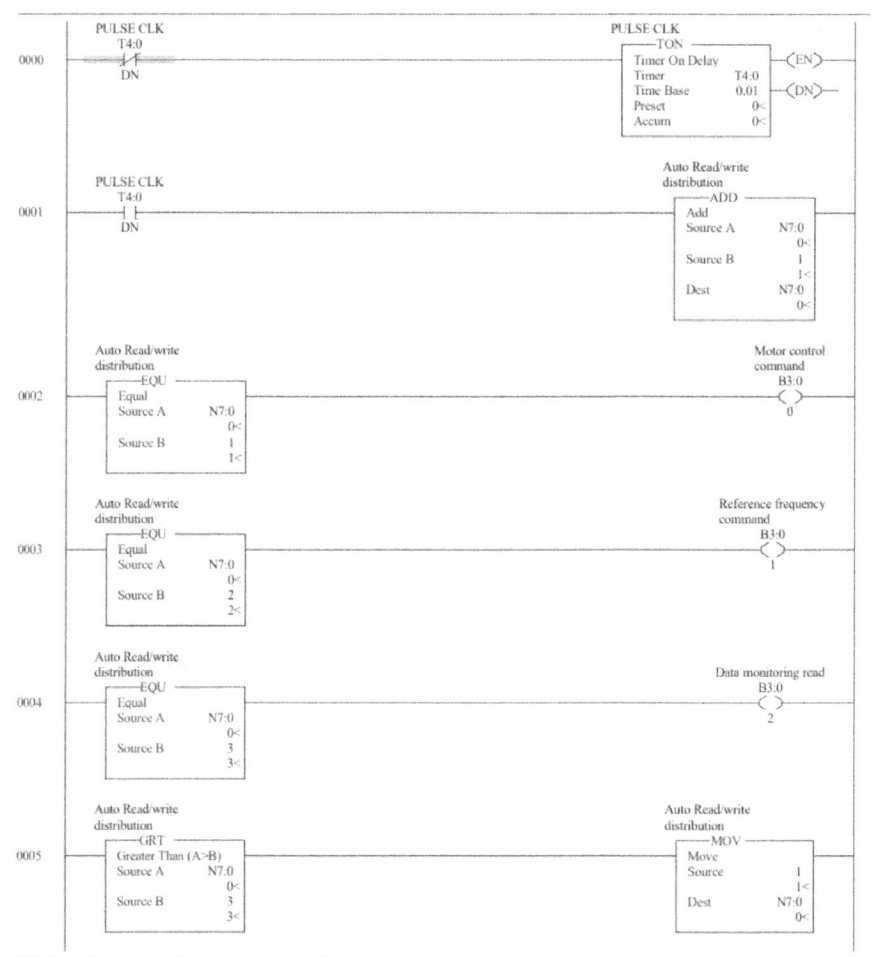

```
0000  PULSE CLK T4:0 DN /                    PULSE CLK T4:0
                                             ---TON---
                                             Timer On Delay        (EN)
                                             Timer         T4:0
                                             Time Base     0.01    (DN)
                                             Preset        0<
                                             Accum         0<

0001  PULSE CLK T4:0 DN                      Auto Read/write distribution
                                             ---ADD---
                                             Add
                                             Source A      N7:0
                                                           0<
                                             Source B      1
                                                           1<
                                             Dest          N7:0
                                                           0<

0002  Auto Read/write distribution           Motor control command
      ---EQU---                              B3:0
      Equal                                  ( )
      Source A      N7:0                     0
                    0<
      Source B      1
                    1<

0003  Auto Read/write distribution           Reference frequency command
      ---EQU---                              B3:0
      Equal                                  ( )
      Source A      N7:0                     1
                    0<
      Source B      2
                    2<

0004  Auto Read/write distribution           Data monitoring read
      ---EQU---                              B3:0
      Equal                                  ( )
      Source A      N7:0                     2
                    0<
      Source B      3
                    3<

0005  Auto Read/write distribution           Auto Read/write distribution
      ---GRT---                              ---MOV---
      Greater Than (A>B)                     Move
      Source A      N7:0                     Source        1
                    0<                                     1<
      Source B      3                        Dest          N7:0
                    3<                                     0<
```

VFD Control Command

VFD Control Address = 2000 (Hexa)
Set Word 2000 = 18 >>> Motor Forward Start
= 34 >>> Motor Reverse Start
= 1 >>> Stop
= 8 >>> VFD Fault Reset
VFD Frequency Reference Address = 2001 (Hexa)

LEARN EVERYTHING ABOUT PLC PROGRAMMING

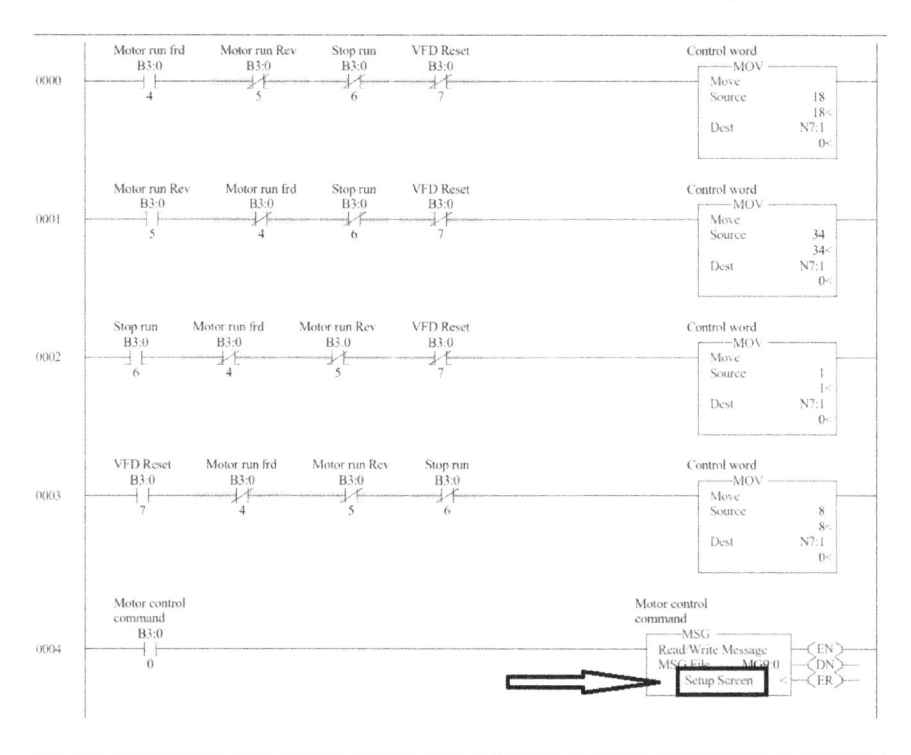

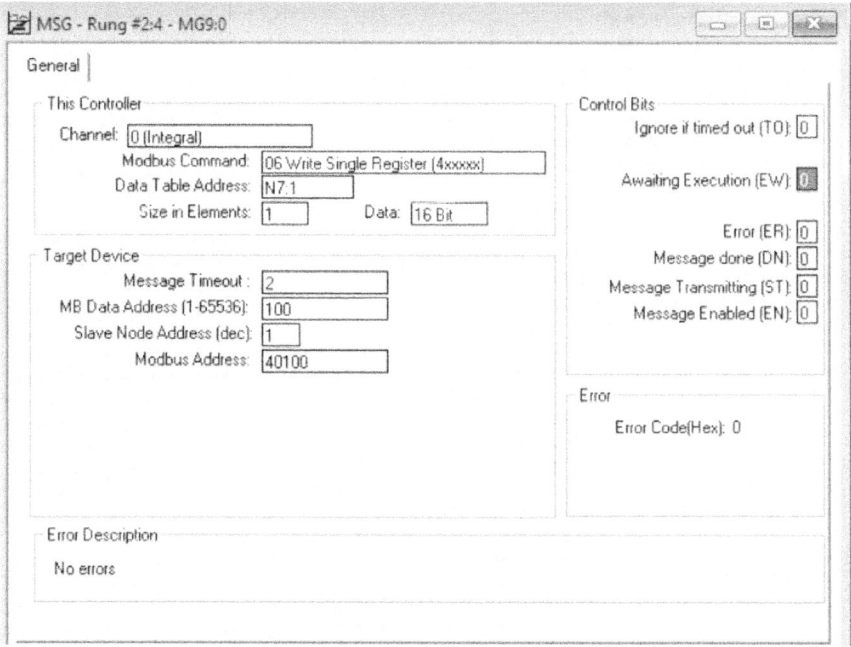

Frequency Setting Command

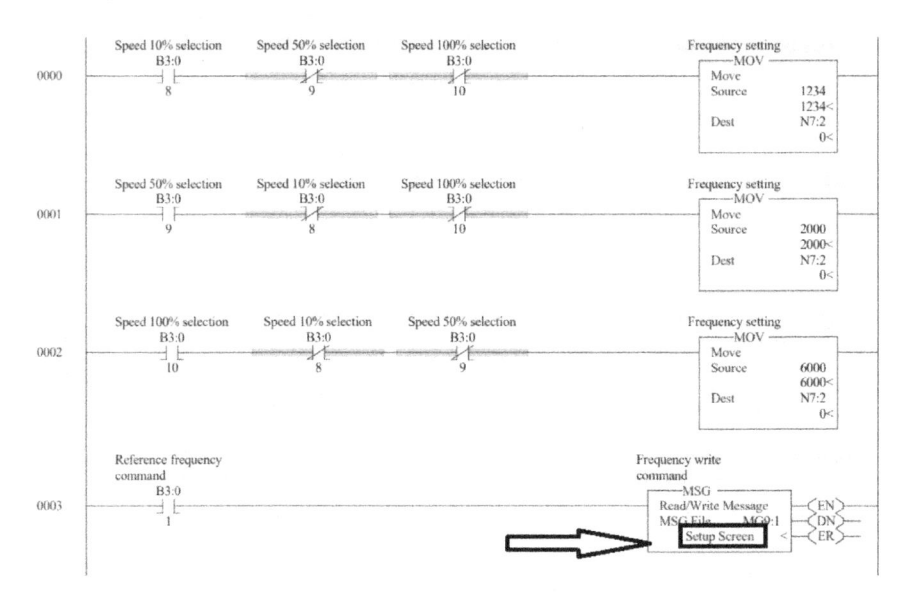

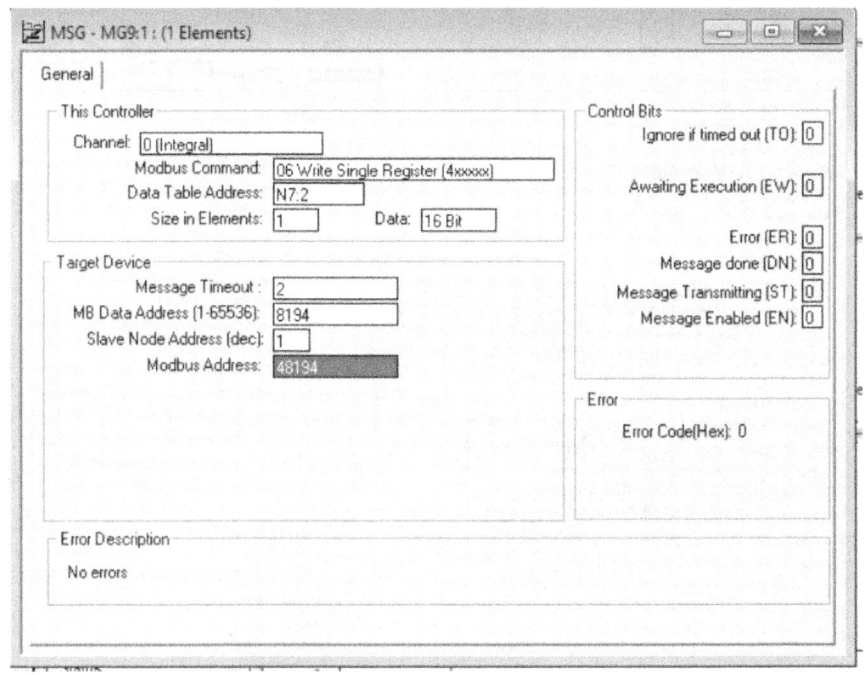

Data Response Monitoring

LEARN EVERYTHING ABOUT PLC PROGRAMMING

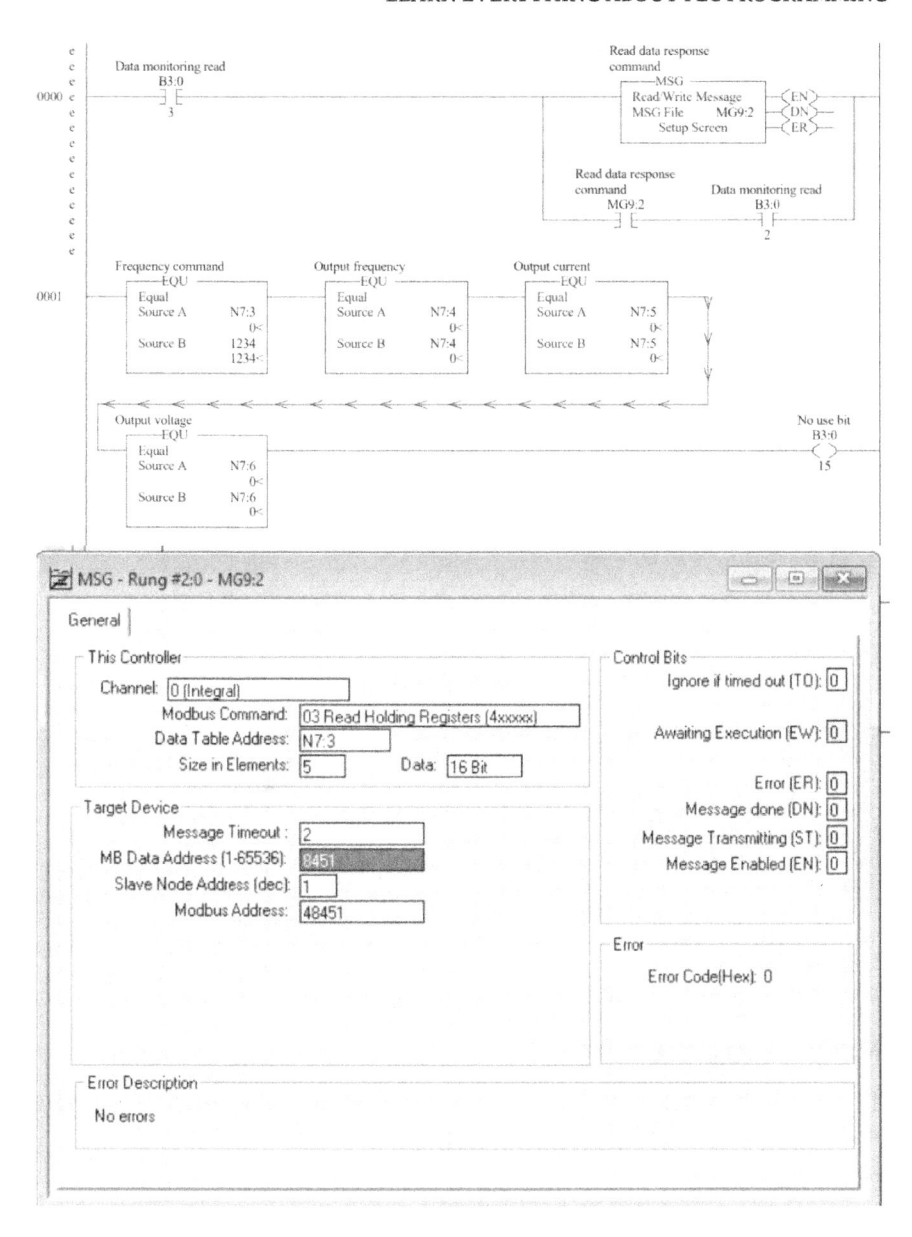

73

CASE STUDY 1

Spray painting application

The conveyor line consists of a spary painting nozzle with a sorting station. The bottles travel over the conveyor with equal distance. Two photo sensors are placed on the conveyor to detect the bottle. The PE 1 sensor detects the bottle, whereas PE 2 Sensor detects only a bigger bottle. A pneumatic rejector is provided to sort the bigger bottle at the end of the conveyor.

Machine Architecture

The entire conveyor line is controlled by PLC and VFD, the details have been mentioned below.
PLC -MicroLogix 1400
VFD - Powerflex 4M (Hard wired control)

Input:

I0:0/0 - Start PB
I0:0/1 - Stop PB
I0:0/2 - VFD status

I0:0/3 - Photocell 1
I0:0/4 - Photocell 2

Output:

O0:0/0 - VFD Run command
O0:0/1 - Spray nozzle
O0:0/2 - Rejection cylinder

Powerflex 4M VFD Parameters:

The PLC controls the VFD via hardwire, and its parameters are as follows:

P106 (Start Source) - 2

P108 (Speed reference) - 4

P109 - Accl Time

P110 - Deccl Time

A410 - Frequency (50Hz)

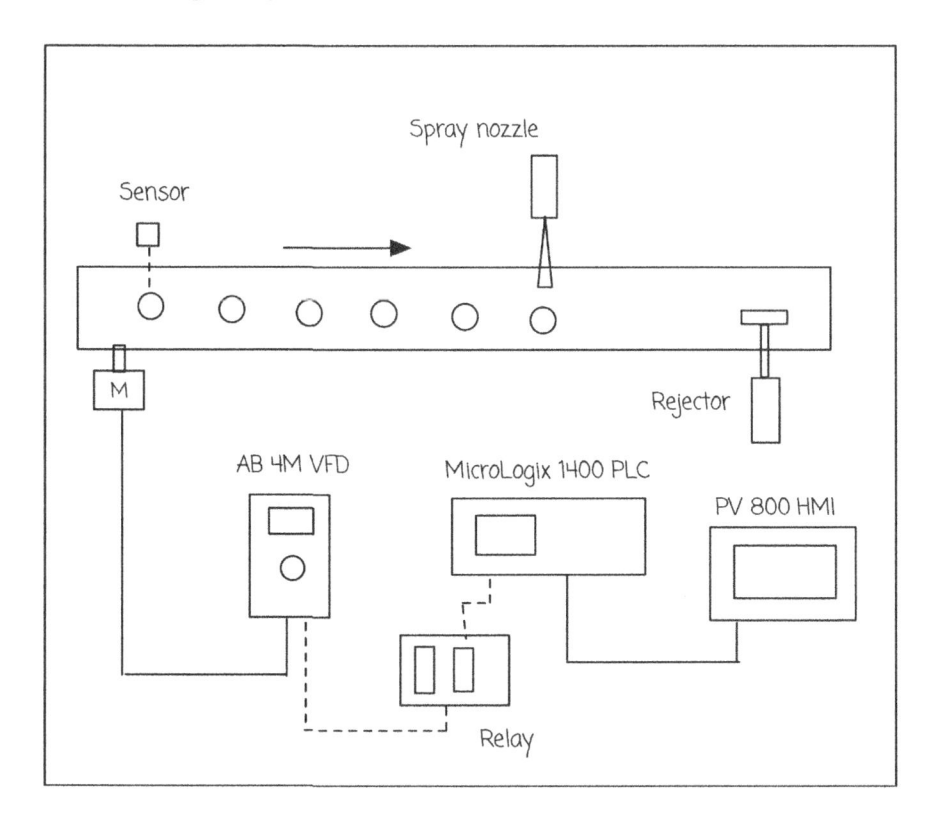

Configuration of PLC
Select the MicroLogix 1400 PLC from the list

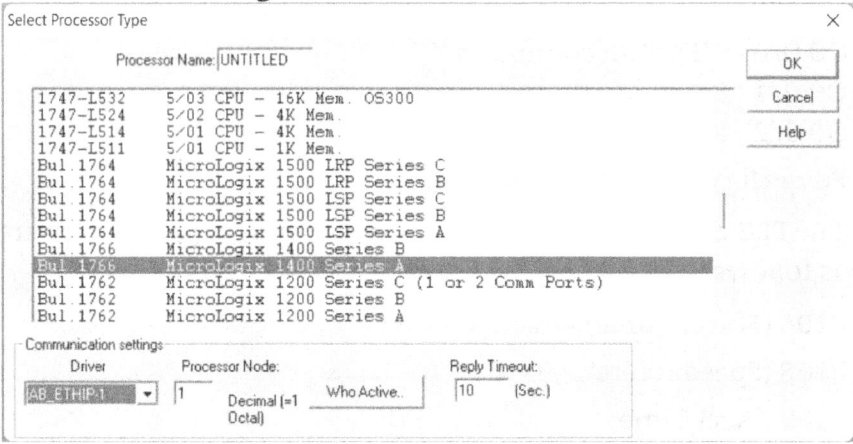

Ladder Logic program

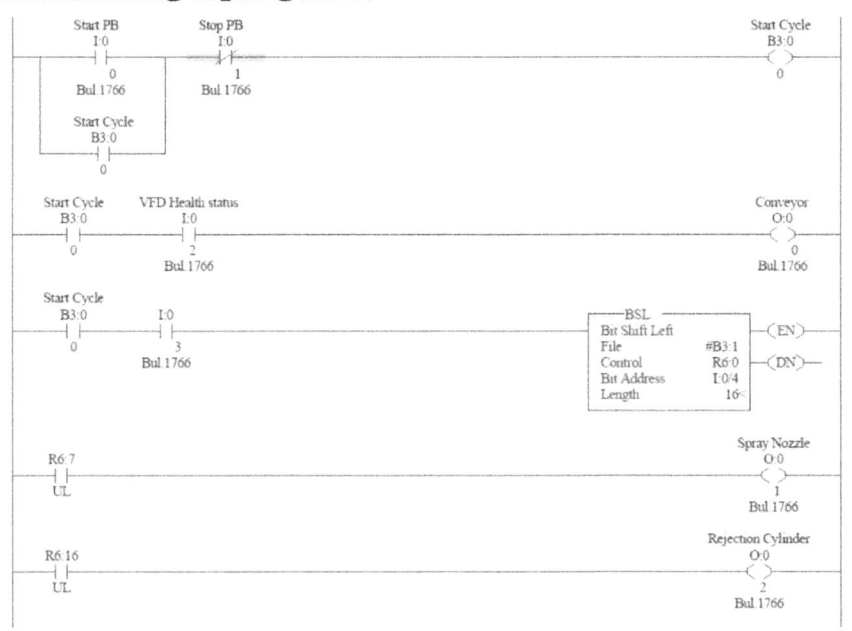

Rung 1: I have used start and stop buttons for a latching circuit.

Rung 2 : The conveyor starts when the latching circut is true and VFD is in healthy state.

Rung 3: I have used a BSL (Bit Shift Register) instruction to make this application successful. But the BSL instruction can be used only when the distance between the bottles remains constant.

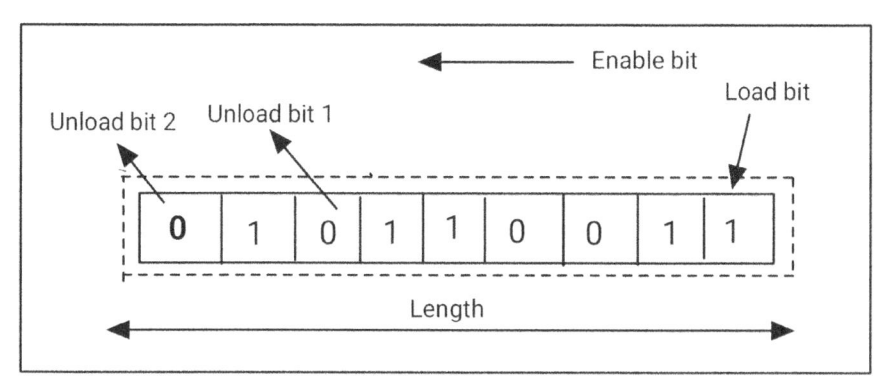

```
                              ← ———— Enable bit
                                            Load bit
Unload bit 2   Unload bit 1
      ↖          ↖                              ↓
   ┌─────┬───┬───┬───┬───┬───┬───┬───┬───┐
   │  0  │ 1 │ 0 │ 1 │ 1 │ 0 │ 0 │ 1 │ 1 │
   └─────┴───┴───┴───┴───┴───┴───┴───┴───┘
   ←——————————————— Length ———————————————→
```

Rung 4 and 5: I have placed a spray nozzle at 7th bit and rejection cylinder at 17th bit of file, so I moniter two bits from the file and unload them from location 7 and 17. As soon as these bits get true, the spray nozzle and rejection cycle perform their work.

2.2 STUDIO 5000 LOGIX DESIGNER

A. Introduction

The Studio 5000 Logix designer is one of the popular software of Rockwell automation. This software supports both ControlLogix and CompactLogix PLC series. It is a complete package of important features such as safety, motion group, and standard program. Also, it supports to third party devices through EDS files which makes it more flexible.

You can download this software from www.rockwellautomation/download.com website and start your journey.

Create a project

Before you can get to the main window for studio 5000 Logix designer you must have a project created. If you don't already have an existing project, click on New project.

A new window will get opened in front of you, where you have to select the type of controller from the list. Select the Revision of software with security authority and select Next.

Software window

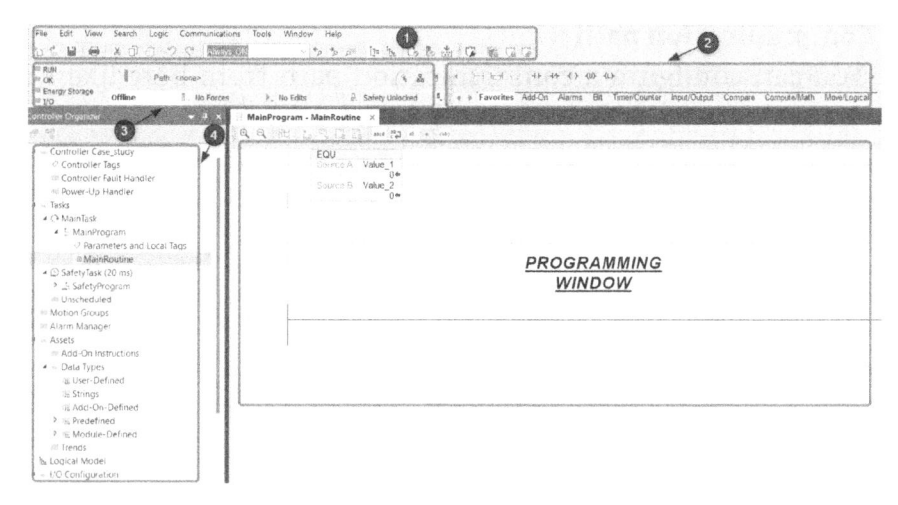

After you finish setting up your project, the actual programming interface window will appear. The window is broken up into few sections. All the toolbars and software navigation are at the top, project navigation is on the left-hand side, the actual programming window is in the middle.

1. Toolbar

It includes following tools:

a. Save and create a program

b. Download and upload a program

c. Verify the program- To check the error in program before downloading

d. Safety mapping (Please refer chapter no. 6)

e. Communication with PLC

f. Monitor tags

g. EDS hardware installation tool- To configure vendor modules

h. Control flash tool- To update software firmware

2. Instruction bar

It is the package of different instructions which are used in a PLC program. Please refer chapter no.4 to learn more.

3. Controller information

Communication path

-You can configure a communication path from here like RS Links.

Controller status

There is different controller status, such as:
- Offline
- Online
- Monitor

Safety Lock/Unlock

- A safety program can be locked and unlocked from here.

4. Project navigation

a. Controller- This includes the information about the controller and controller tags that can be accessed by standard and safety programs.

b. Task - It provides scheduling and priority information for set of one or more programs.

c. Program- Each task requires at least one program that defines the operations.

d. Routine - It provides executable code for the project in a controller. Each routine uses a specific programming language (Ladder, Functional block dia, Sequential Flowchart, or structured text)

e. Main Routine -The first routine executed in a program. Use the main routine to call other routines using jump to subroutine (JSR) instruction.

f. Subroutine - Any routine other than subroutine

g. Assets - It is like a box of treasure, it includes the user created add on instructions that can be used again and again.

B. Getting started with Studio 5000 Logix designer

A. Set up a controller or module

When you write a program from the scratch, or sometimes

you need to replace a module due to the failure in machine. So in both cases you have to follow certain steps in order to work modules smoothly.

Step 1- Set IP address

Each module has a different way of setting IP address, the method of a few modules have been explained as below.

Controller: When you power on the controller, the first step is to set the IP address of the controller using any one of the methods explained as above. Personally I use Bootp DHCP method to set the IP address.

VFD: In case of Powerflex 525 VFD, you have to set the IP address by changing the communication parameters.

AENT (Ethernet Adapter): In this case, you have to set the last node of IP address on AENT module. For ex. 198.168.1.3 you have to set lase node '3' on the AENT module.

Armor block: You have to set the last node of IP address by rotating screws given on the module.

Step 2- Check the module configuration of the controller using RS Linx to know about the firmware revision of the controller.

Step 3- Update firmware

Download the firmware from the Rockwell website and update the firmware using ControlFlash software. The extension file of firmware is .dmk file.

Step 4- Configuration using studio 5000

Whenever you are going to configure a module in studio 5000 Logix designer make sure the module configuration shown in the RS Linx is same as you are entering in studio 5000.

Now your controller or module is ready to use.

The next step after communication is configuration of peripheral devices. In this chapter, we are going to learn about configuration of new modules, setting up IP their addresses, and communicating with them using IP addresses.

1. Configuration of Allen Bradley module

Suppose, you have physically connected a module with PLC, but yet to be configured in studio 5000 software, then a yellow triangle will appear over the module when you go online and also PLC's IO LED will start blinking. So, first it is necessary to properly configure a device in studio 5000 Logix designer.

After configuration, the controller tags get generated, through which you can control a module. Here is the list of Allen Bradley devices that can be configured.

AB Modules

1. Communication- AENT module
2. Digital and analog IO cards
3. Safety IO cards
4. Power flex VFD ethernet embedded
5. Motion- Servo drive
6. Controller
7. Encoder
8. HMI
9. Energy management devices
10. Motor starter and overload

B. Configuration of Remote IO module

In this section, I am going to show you the configuration of Ethernet remote IO module with different input and output cards.

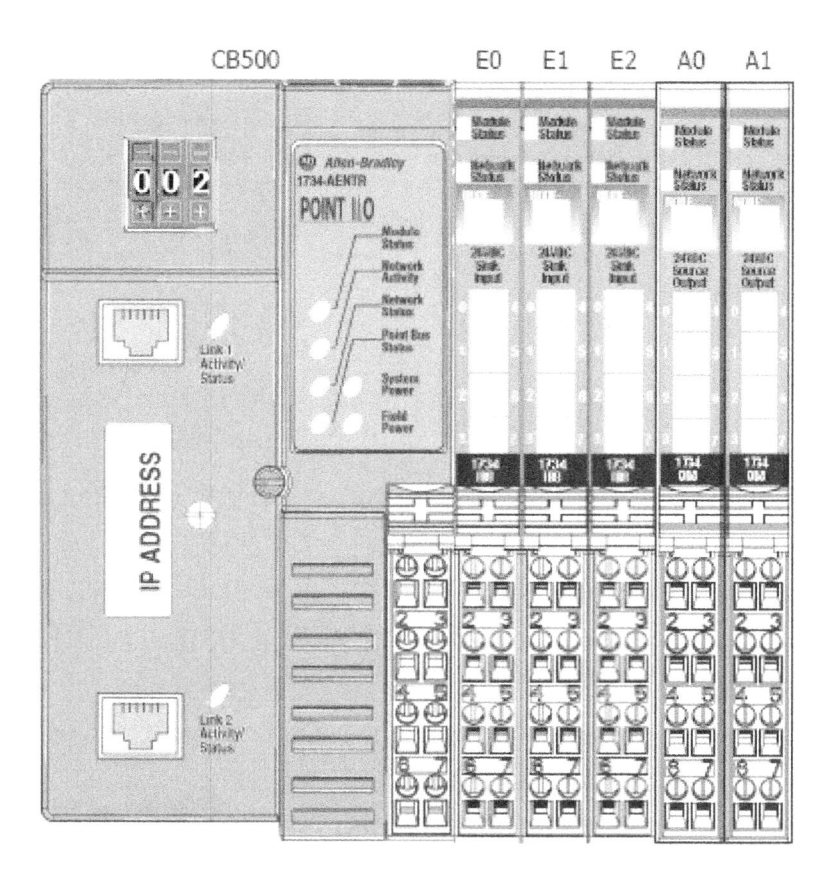

Follow the steps given below to add an ethernet module in your program.

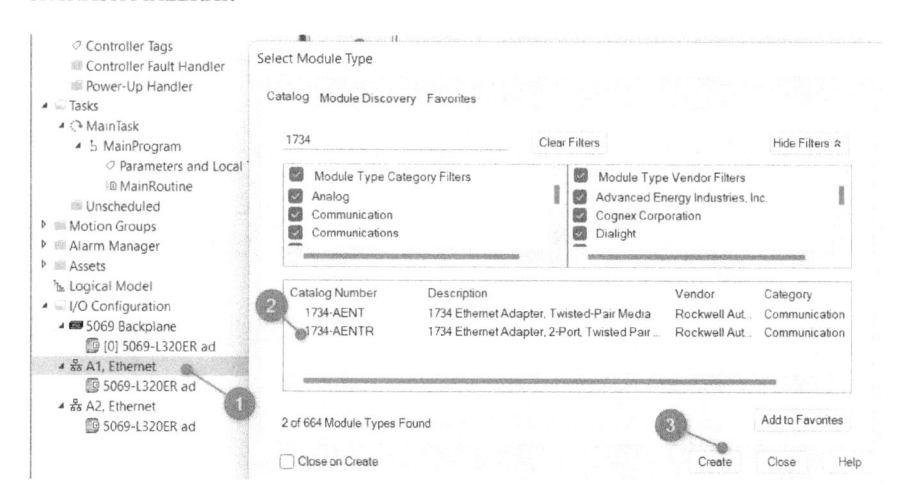

1. Right click on Ethernet A1, and select add module.
2. Select 1734-AENTR communication adapter from the list.
3. Click on create
4. Name the module as you want.
5. Set the IP address of modules that must be in series with IP address of PLC.
6. You can increase or decrease the size of chassis from here.
7. Select the size of chassis and click OK.

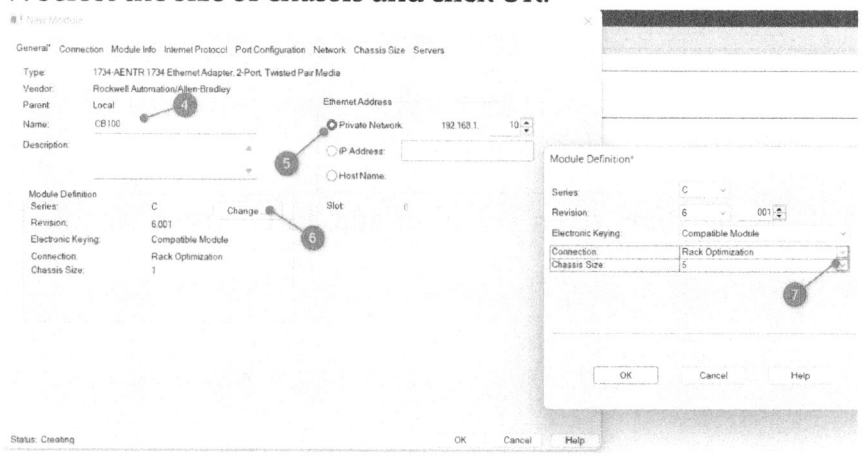

Electronic keying

The electronic keying is like providing a lock to the module in a software.

1. Exact match- It is used to match minor and major revisions of a module.

For example, an Armor block 1732E-CFGM12R series A with revision 1.008 can be replaced by 1732E-CFGM12R series A with revision 1.002.

2. Compatible module- This type of keying is used to match different revisions of a module.
For example, Armor block 1732E-CFGM12R series A with revision 1.008 can be replaced by 1732E-CFGM12R series A with revision 3.002.

3. Disable keying- It is used to disable the keying function of a module. You can provide any third party module which should serve the same purpose.

For example, 1732E-CFGM12R series A can be replaced with 1732E-CFGM12R series B.

Configuration of IO cards

Follow the following steps:
1. Select the chassis you want to configure and right click to add module.
2. Select the digital input IB8.
3. Name the card as you want.
4. Choose the slot of chassis where IO module is fixed and click on 'OK'.

Similarly, follow the same steps to configure I01, I02, Q00, and Q01.

2. Configuration of vendor module

This technique is used to add devices which don't belong to Rockwell automation family. In order to add such a device, you need to provide their EDS file to studio 5000 Logix designer.

- Using EDS file
- Using Generic ethernet module

A. Configure a device by importing EDS file

The Electronics Data Sheet (EDS) files are simple text files used by network configuration tools to help you identify products and easily commission them on a network.

The EDS files can be imported by two methods. But for that you need to have preinstalled software in your PC. You need to have an 'EDS hardware installation tool' which is already installed with studio 5000 Logix designer.

Follow the following steps:

Step 1: Search 'EDS hardware installation tool' in search box of window.

This software is already installed with studio 5000 Logix designer, so you no need to put extra efforts.

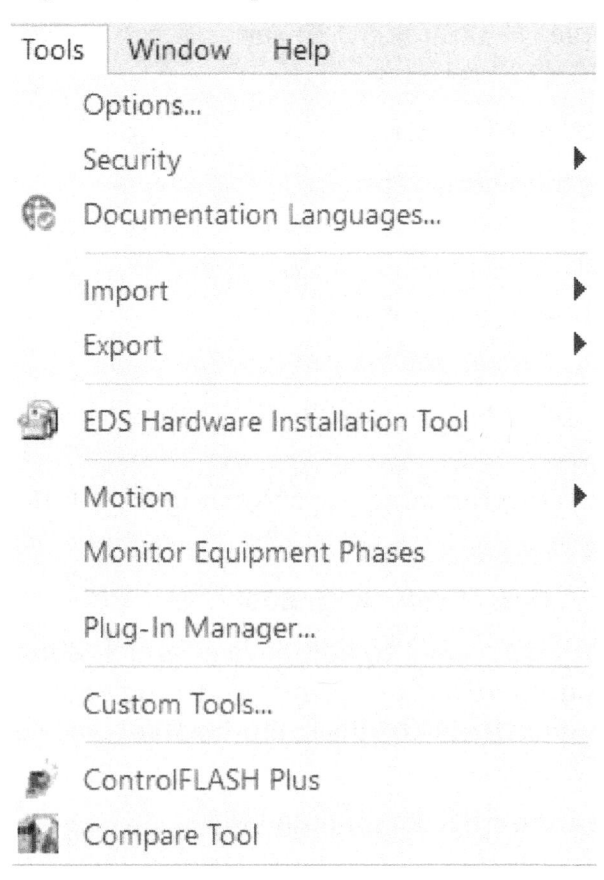

Otherwise, you can find this tool from tool option in studio 5000 Logix designer. When you open the EDS wizard, you see two options to add or remove EDS files.

Rockwell Automation - Hardware Installation Tool 33.0.20.0

This tool allows you to change the hardware description information currently installed on your computer.

| Add | Launch the EDS Wizard and add selected hardware description files only. |
| Remove | Launch the EDS Wizard and remove selected hardware description files only. |

Exit

Step 2:

For the demonstration purpose, I have chosen a camera for vision system. When I have installed insight explorer software, its EDS files got stored in C drive of computer by default.

When I click to add button to add an EDS file, the software asks me to browse the path of EDS files.

Step 3:

As you can see here, all the EDS files have got an extension as .eds

After importing an EDS file, the particular module will be visible in the list of modules.

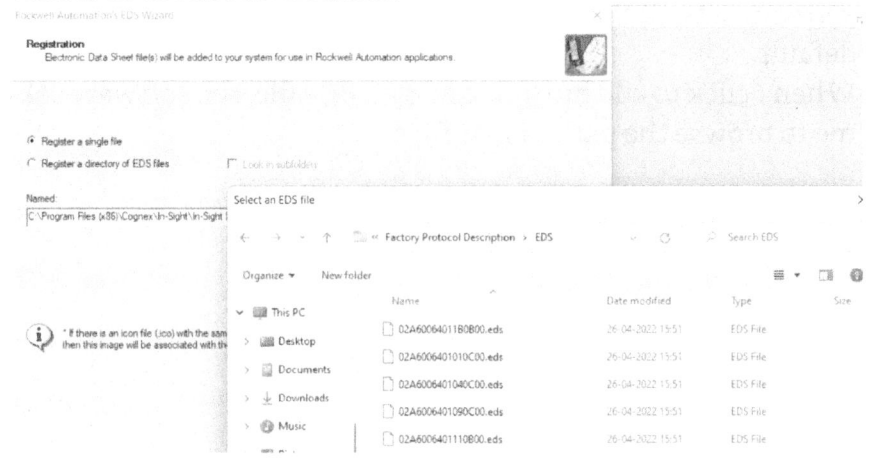

Step 4:

If there is a valid EDS file exists in that folder, a green tick will appear on the file.

Step 5:

Now, you can see cognex module on the screen.

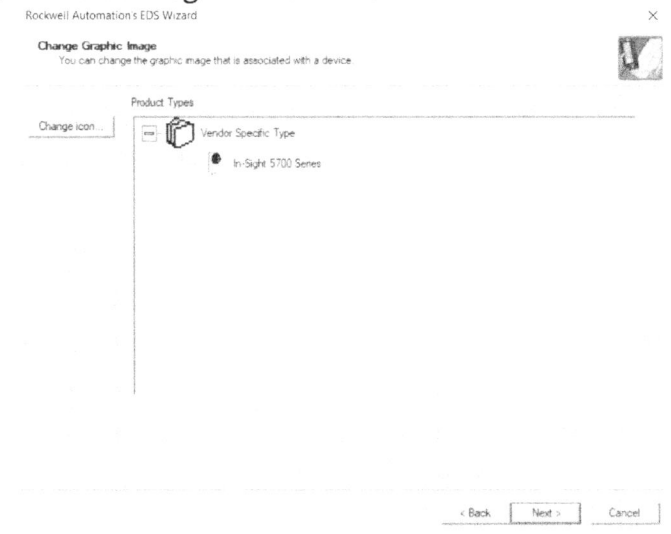

Step 6:

After importing an EDS file, the particular module will be visible in the list of modules.

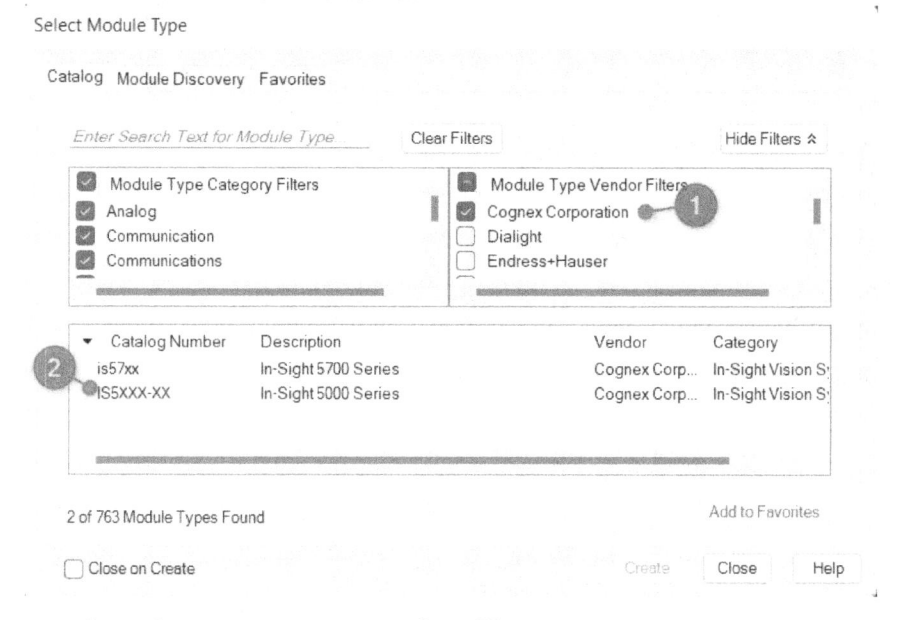

1. Select the Cognex corporation filter.
2. Select the module type and click on 'OK'.

B. Configuration using Generic ethernet module

A Generic Ethernet module is used when the vendor modules don't have EDS file. The studio 5000 Logix designer has a provision to set a communication between such a module by using Generic ethernet modules.

LEARN EVERYTHING ABOUT PLC PROGRAMMING

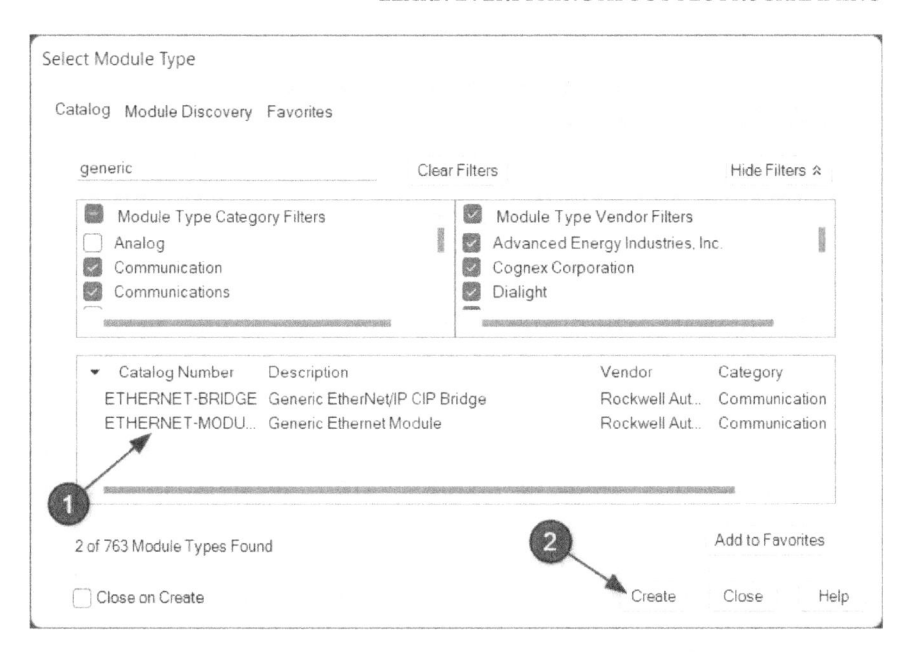

1. Right click on the A1 ethernet, click on add module, and select Generic ethernet module.

2. Click on create.

A new window will appear.

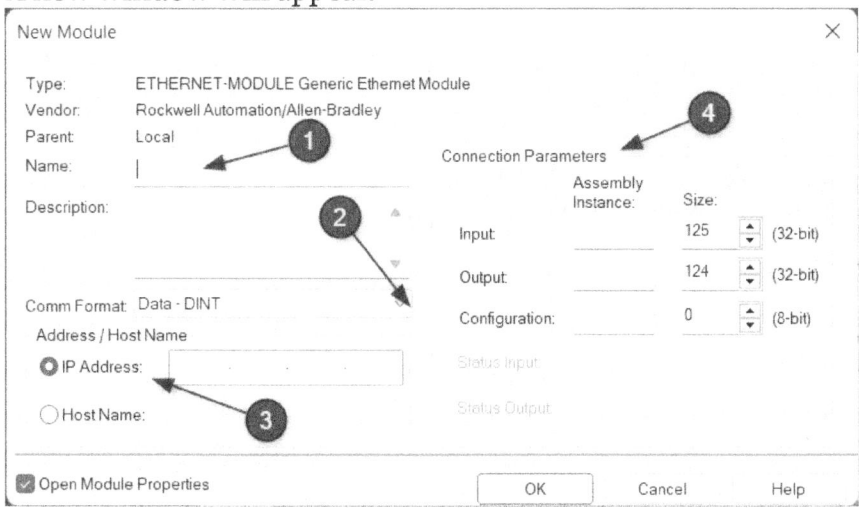

1. Name the module as you want
2. Select communication format from the list with data type.
3. Set IP address

91

4. Set connection parameters and click OK

C. Update firmware

The **firmware** is a type of software that is embedded in hardware devices. It is responsible for controlling the operation of the device and ensuring that it functions correctly.

The firmware is programmed in a high level language, such as C or C++, and it is then compiled into the machine code that can be executed by the device's processor. It communicates with other hardware components in the device to perform various fuctions, such as reading input signals, processing data, and controlling output signals.

Procedure to update firmware:

1. Download latest firmware from www.rockwellautomatio.com/download.

2. Copy .dmk file and paste it in program data(x86)< controlFlash software.

3. Connect the module to your PC with Ethernet or USB cable.

4. Open ControlFlash software.

5. If your computer has firmware file, it can be seen in the list.

6. Select the module from the list.

7. Now you can see the current firmware of the module and the list of firmware that your computer has.

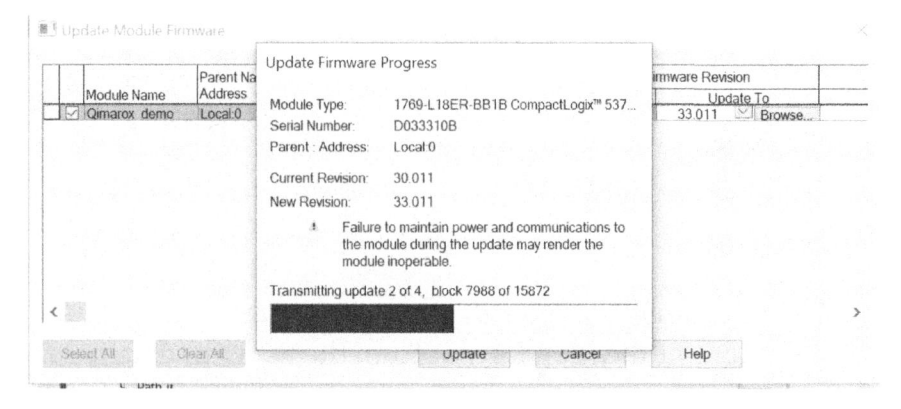

8. Click on update.

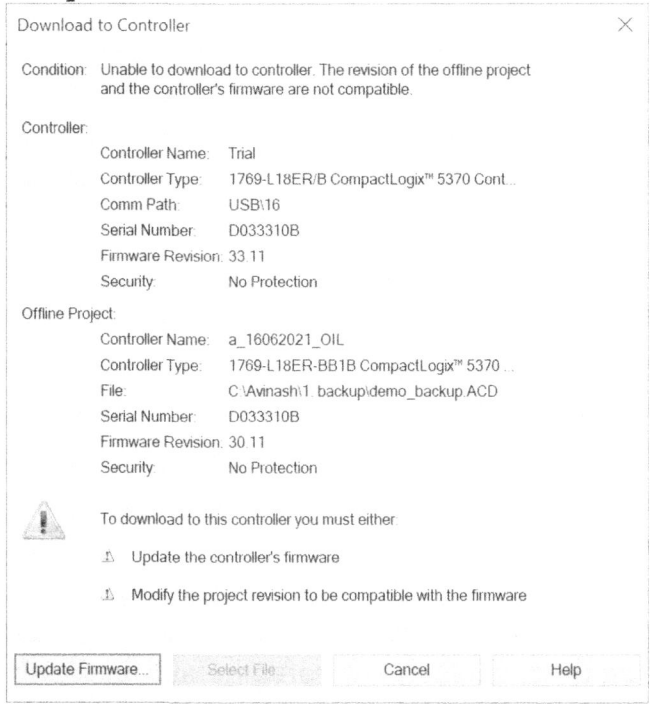

> **Download to Controller**
>
> **Condition:** Unable to download to controller. The revision of the offline project and the controller's firmware are not compatible.
>
> **Controller:**
> - Controller Name: Trial
> - Controller Type: 1769-L18ER/B CompactLogix™ 5370 Cont...
> - Comm Path: USB\16
> - Serial Number: D033310B
> - Firmware Revision: 33.11
> - Security: No Protection
>
> **Offline Project:**
> - Controller Name: a_16062021_OIL
> - Controller Type: 1769-L18ER-BB1B CompactLogix™ 5370...
> - File: C:\Avinash\1_backup\demo_backup.ACD
> - Serial Number: D033310B
> - Firmware Revision: 30.11
> - Security: No Protection
>
> ⚠ To download to this controller you must either:
> - Update the controller's firmware
> - Modify the project revision to be compatible with the firmware
>
> [Update Firmware] [Select File] [Cancel] [Help]

Note: Make sure the continuity of power supply during firmware upgradation.

Update controller firmware

The firmware major revision level must match the software major version level. For example, if the controller firmware revision is 31.xxx, you must use the Logix Designer application, version 31.

I have considered an example of a controller which as a program developed in studio 5000 Logix designer V30 and I have a software which has V33. In order to upload or download a program, I have to match the software revision. To update your controller firmware with the Auto Flash feature, complete these steps.

1. Verify that the network connection is made and your

network driver is configured in Linx-based communication software.

2. Use the Logix Designer application to create a controller project.

3. On the Path bar, click Who Active.

4. Click on download.

5. On the Who Active dialog box, select your controller under the communication driver you want to use, and click Update Firmware.

D. IO cards addressing

I/O information is presented as a set of tags.
Each tag uses a structure of data. The structure depends on the specific features of the I/O module.
The name of the tag is based on the location of the I/O module in the system. When you add a module to the I/O Configuration folder, the software automatically creates controller-scoped tags for the module in Controller Tags.
An I/O address uses this format: **Local:2:I.Data.3**

Tagging

In simple words, tagging is renaming of any devices or instructions. For example, a PLC address I0:0/0 can be renamed as a start push button. Doesn't is sound easier than recalling by its address?

Creating new tags

Creating a tag during programming is one of the easier methods. To create a tag, right click on the tag and select 'new tag' and follow the following procedure.

1. Name the tag with the description
2. Select the usage of tag, the usage could be:
-Local tag
-Input Parameter
-Output Parameter

-InOut Parameter
-Public parameter
3. Tag Types
Base- These tags store values for use by logic within the project.
Alias- A tag that references another tag. An alias tag can refer to another alias tag or a base tag. An alias tag can also refer to a component of another tag by referencing a member of a structure, an array element, or a bit within a tag or member.
IMPORTANT: Aliasing between standard and safety tags is prohibited in safety applications. Instead, standard tags can be mapped to safety tags using safety tag mapping.
Produced- A tag that a controller makes available for use by other controllers. A maximum of 15 controllers can simultaneously consume (receive) the data. A produced tag sends its data to one or more consuming tags without using logic. Produced tag data is sent at the RPI of the consuming tag.
Consumed- A tag that receives the data of a produced tag. The data type of the consumed tag must match the data type of the produced tag. The requested packet interval (RPI) of the consumed tag determines the period at which the data updates.
4. Data Type- The data types could be Bool, Timer, Counter, etc.
5. Scope- The defined tag could be a tag of main program or of safety program.
6. External Access- It allows to edit the tag or disable editing option.
7. Constant
8. Click on 'Create'.

E. Controller and Local tags

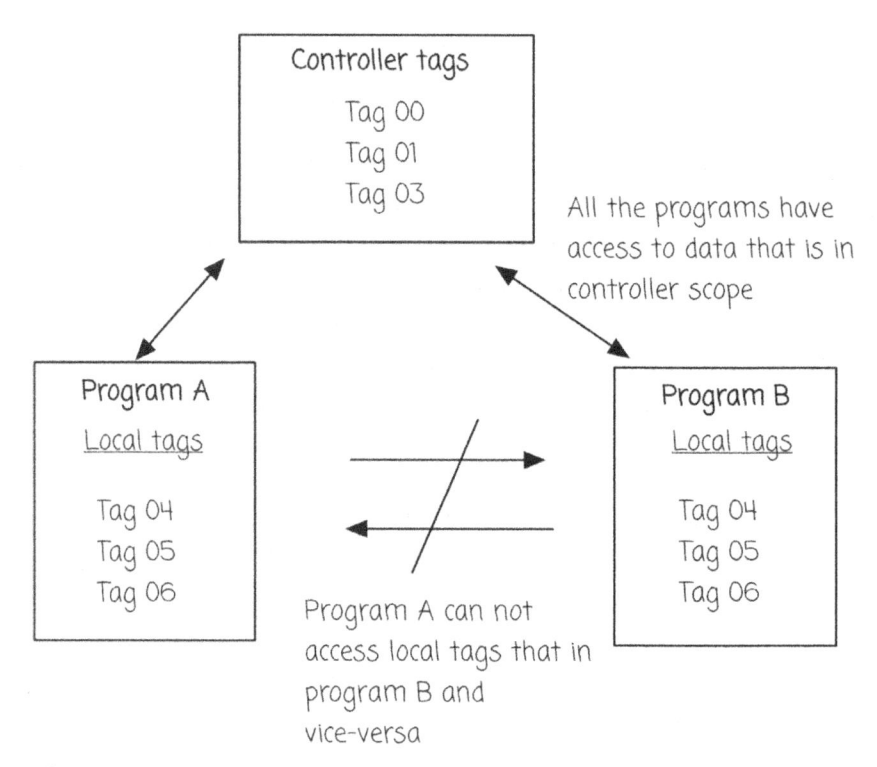

```
         ┌─────────────────┐
         │  Controller tags│
         │     Tag 00      │
         │     Tag 01      │
         │     Tag 03      │
         └─────────────────┘     All the programs have
                                 access to data that is in
                                 controller scope
```

┌──────────────┐   ┌──────────────┐
│ Program A   │   │ Program B   │
│ Local tags │   │ Local tags │
│   │   │   │
│   Tag 04   │   │   Tag 04   │
│   Tag 05   │   │   Tag 05   │
│   Tag 06   │   │   Tag 06   │
└──────────────┘   └──────────────┘

Program A can not access local tags that in program B and vice-versa

When you create a tag, you define it as either a controller tag (global data) or a local tag for a specific program (local data).

A Logix 5000 controller lets you divide your application into multiple programs, each with its own data. There is no need to manage conflicting local tag names between programs. This makes it easier to reuse both code and tag names in multiple programs.

Data at the program scope is isolated from other programs.

- Routines cannot access data that is at the local scope (local tag) of another program.

- You can reuse the tag name of a local tag in multiple programs.

For example, both Program_A and Program_B can have a local tag named Tag_4.

- You can also use program parameters to share data between programs as an alternative to controller-scope tags. See Program parameter scope.

F. Routines

Till now you have learned about creating tags, now you need to build the program and routines to use those tags. The routines are used to divide a program into the several parts. This makes a program easier and better for troubleshooting.

Create a routine

You can add as many subroutines as you need to your program, as long as your PLC doesn't run out of memory.
- Right click on the main program, and select add.
- Click on 'New Routine'.

JSR for routine

Your job is not finished yet until you use JSR instruction. If you create a new routine it will not work right away. Only the main routine is executed by default from the controller. Any code you enter in the main routine will execute, but if you create an additional subroutine you will have to call that routine first.

Hence you need to use a JSR instruction in the main program. This instruction jumps from the main program to whichever subroutine you call.

Example:

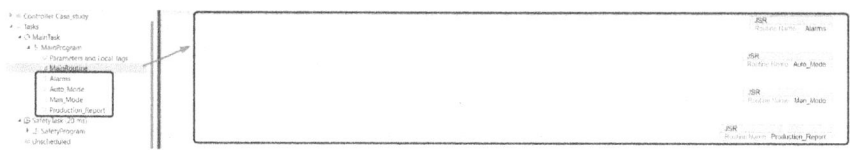

As you can see in the above image, I have already created subroutines below the main program. To execute these subroutines, I have used JSR instructions to call these subroutines.

Rung comments

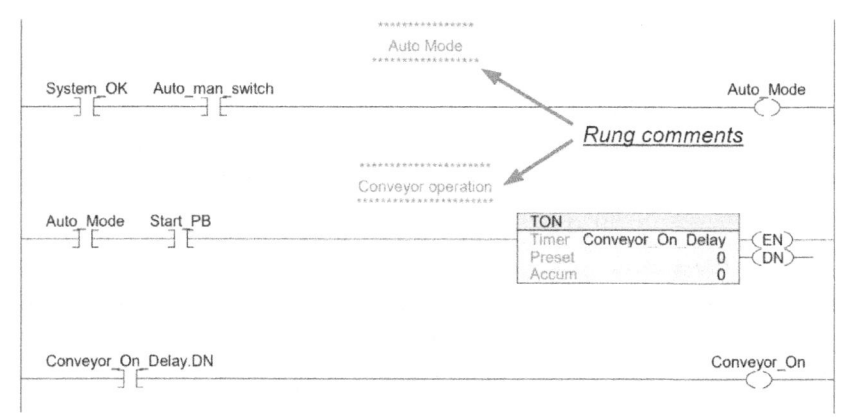

Having a rung commented PLC program is better for troubleshooting and understanding a logic. A programmer should always focus of making a program easier to understand to other by implementing such a practice.

Steps to add a rung comment

- Right click on the rung you wish to add a comment on
- Select 'Edit rung comment' and write a comment.

G. IO buffering or Mapping

IO mapping is a second step of PLC programming after configuration of different devices. It is a step to understand a program. IO Mapping or Buffering can help you in multiple ways. It allows one to contain all the primary input & output tags within a single program & easily manipulate them as needed.

```
▲ Controller ad
    ◇ Controller Tags
    ▩ Controller Fault Handler
    ▩ Power-Up Handler
▲ Tasks
    ▲ ⟲ MainTask
        ▷ ┗ MainProgram
        ▩ Unscheduled
▷ ▩ Motion Groups
▷ ▩ Alarm Manager
▷ ▩ Assets
    ▥ Logical Model
▲ I/O Configuration
    ▲ ▦ 5069 Backplane
        ▤ [0] 5069-L320ER ad
        ▲ ⛬ A1, Ethernet
            ▤ 5069-L320ER ad
            ▲ ▯ 1734-AENT/C C100
                ▲ ▦ PointIO 5 Slot Chassis
                    ▯ [0] 1734-AENT/C C100
                    ▯ [1] 1734-IB8/C IN1
                    ▯ [2] 1734-IB8/C IN2
                    ▯ [3] 1734-IB8/C IN3
        ▲ ⛬ A2, Ethernet
            ▤ 5069-L320ER ad
```

The necessity to manipulate may come from hardware failure or need to upgrade or expand the system. Furthermore, having these assets within a single program greatly reduces the time to troubleshoot and commission new systems.

IO mapping can be done in two ways:
- Move data from IO cards to defined SINT data type
- Move data from one Bool to another Bool

1. Move data from IO cards to defined SINT data type

As you can see in the above image, I have already configured 3 digital input cards named as IN1, IN2, and IN3. Each digital input card has 8 digital inputs. Each digital input has a value of one bit. Therefore, each card has data which is equal to 8 bits, in other word short integer (SINT).

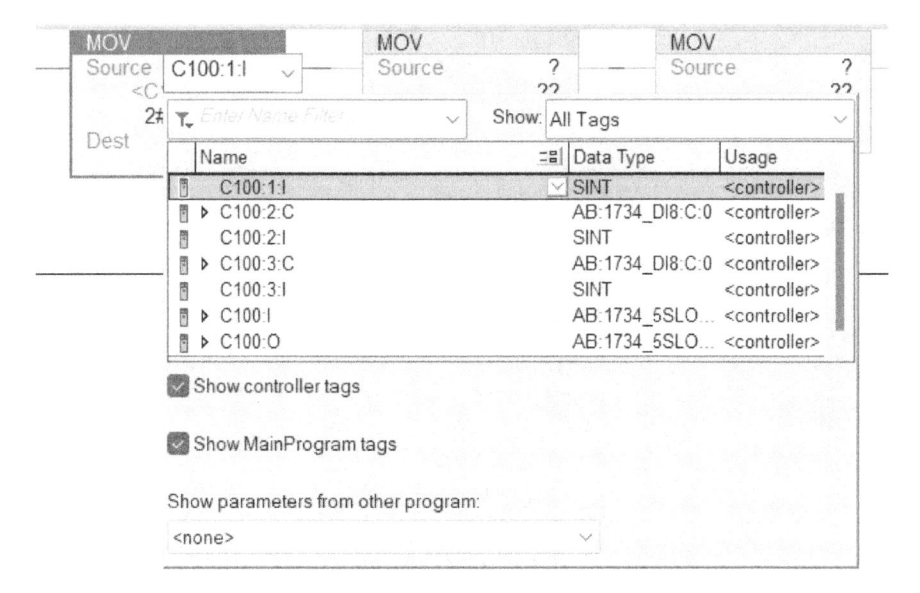

[Screenshot of RSLogix tag browser showing MOV instructions with Source C100:1:I and tag list including C100:1:I (SINT), C100:2:C, C100:2:I, C100:3:C, C100:3:I, C100:I, C100:O]

I have used a move instruction to move data stored in the digital input card which is in the 1st slot of AENT card to the short integer INPUT1. Similarly, for the data stored in the digital input card in the 2nd and 3rd slots of AENT card.

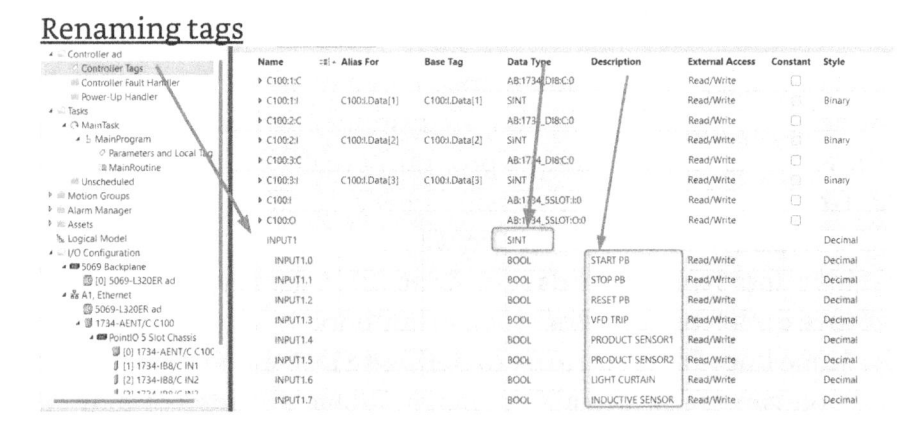

Renaming tags

[Screenshot of Controller Organizer and tag table showing aliases: C100:1:I, C100:1:1 (alias C100:I.Data[1], base C100:I.Data[1], SINT), C100:2:C, C100:2:I (alias C100:I.Data[2], base C100:I.Data[2], SINT), C100:3:C, C100:3:I (alias C100:I.Data[3], base C100:I.Data[3], SINT), C100:I, C100:O, INPUT1 (SINT), INPUT1.0 (BOOL, START PB), INPUT1.1 (BOOL, STOP PB), INPUT1.2 (BOOL, RESET PB), INPUT1.3 (BOOL, VFD TRIP), INPUT1.4 (BOOL, PRODUCT SENSOR1), INPUT1.5 (BOOL, PRODUCT SENSOR2), INPUT1.6 (BOOL, LIGHT CURTAIN), INPUT1.7 (BOOL, INDUCTIVE SENSOR)]

As you can see in the above image, you have to first create a short integer in the controller tag 'Input1', and then assign a particular bool the different inputs devices such as push buttons and sensors.

This method can be used when there are number of IO cards to configure.

2. Move data from one Bool to another Bool

As you can see, the data moves from the one bool of IO card to the defined Bool. This method can be used when there are a few numbers of IO cards to configure.

H. Program Upload and Download

1. Upload a program

When you open studio 5000 software, the screen will open in front of you. You can select 'from upload' option to upload a program from a controller to the computer.

1. Select communication software
2. Select the controller
3. Click upload

2. Download a program

Follow these steps to transfer your project from your computer to your controller.

1. Turn the key switch of the controller to REM.
2. Open the RS Logix 5000 project that you want to download.
3. Define the path to the controller.

a. Click Who Active.

b. Select the controller.

To open a level, click the + sign. If a controller is already selected, make sure that it is the correct controller.

4. Click Download.

The software compares the following information in the offline project and the controller:

- Controller serial number (if project to controller match is selected)
- Firmware major and minor revisions
- Safety status
- Safety task signature (if one exists)

- Safety-lock status

I. Ladder Logic Instructions

1. Bit Instructions

These bit instructions are used for digital signal programming.

XIC (Examine if Closed)

This instruction allows a program to scan if the bit is true of equal to 1. In short we call it NO contact.

XIO (Examine of Open)

This instruction is opposite to XIC and it allows a program to scan if the bit is false of equal to 0. In short we call it NC contact.

OTE (Output Energize)

This OTE coil gets energize if the previous instructions are true.

OTL (Output Latch)

This instruction will get change permanently from 0 to 1, if all the previous instructions are true for once. This instruction is like a latching or holding circuit.

OTU (Output Unlatch)

This instruction will get change permanently from 1 to 0, if all the previous instructions are true for once.

ONS (One Shot)

Let's understand this instruction with a simple example. As you can see in the above image, a part present sensor detects the presence of a bottle and counts the bottle in an incremental manner.

The counter only counts the pulses received from sensor. I have used an ONS instruction here to avoid any miscounting.

2. Timer

The timer instruction plays a crucial role in machine's operation. We come across few applications where we need to

delay the operation or perform an operation within specific timer period. In the both cases, we require a timer. Studio 5000 Logix designer has three types of timer.
- TON
- TOF
- RTO

a. TON (Timer On Delay)

```
TON
Timer           Timer      (EN)
Preset             0◄      (DN)
Accum              0◄
```

When the start PB is set, red lamp is on for 1000 milliseconds, that means Timer.TT bit will be true for same timer period. Once the timer reaches 1000 Ms, red lamp will get off, and green lamp will glow.

Like wise, this type of timer is used to switch on a particular output after set time.

b. TOF (Timer Off Delay)

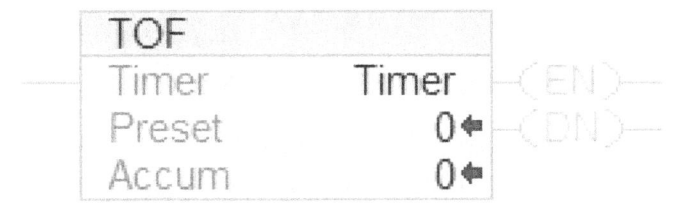

```
TOF
Timer           Timer      (EN)
Preset             0◄      (DN)
Accum              0◄
```

This type of timer requires just a pulse of input to start. When the timer gets one shot, the Timer.TT bit will get on for 5 seconds. This timer does not require continuous supply of bit.

c. RTO (Retentive Timer On Delay)

```
RTO
Timer          Timer   —(EN)—
Preset            0←   —(DN)—
Accum             0←
```

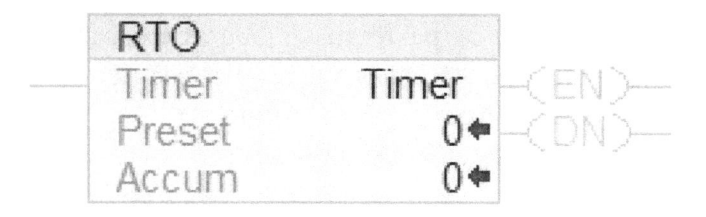

When the start PB is true, green lamp gets on for 5000 milliseconds. When timer.acc reaches 5000, green lamp goes off and red lamp goes on. The red lamp remains on until timer is reset. If start PB is cleared while timer is timing, green lamp goes off (that means timer needs a continuous high bit). When reset PB is set, the RES instruction resets timer (clears status bits and .ACC value).

3. Counter

The counter is used to count the number of pulses. It counts the pulses in incremental or decremental manner. The studio 5000 software offers two types of counter.

a. CTU (Count Up)

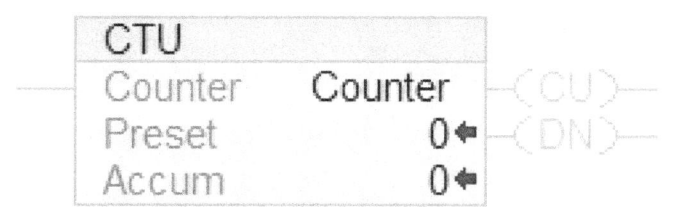

```
CTU
Counter        Counter  —(CU)—
Preset            0←    —(DN)—
Accum             0←
```

- The CTU counter counts the false to true transition in increment manner and stores the count value in the Accum.
- The output of the counter gets true when the Preset value = Accum value.
- This counter requires Reset coil to reset the counting value.

b. CTD (Count Down)

```
CTD
Counter    Counter
Preset          0
Accum           0
```

- The CTU counter counts the false to true transition in decrement manner and stores the count value in the Accum.
- The output of the counter gets true when the Preset value = Accum value.
- This counter requires Reset coil to reset the counting value.

4. Input/Output instructions

The input/output instructions read or write data to or from the controller or a block of data to or from another module on another network.

- GSV/SSV
- MSG
- IOT

a. GSV (Get System Value) and SSV (Set System Value)

```
GSV                                SSV
Class Name     WallClockTime       Class Name     WallClockTime
Instance Name                      Instance Name
Attribute Name   DateTime          Attribute Name  LocalDateTime
Dest                 Time          Source              Time
                        0                                  0
```

The GSV/SSV instructions get and set controller status data that is stored in objects. The controller stores status data in objects.

- When true, the GSV instruction retrieves the specified information and places it in the destination.
- When true, the SSV instruction sets the specified attribute with data from the source.

When you enter a GSV/SSV instruction, the programming software displays the valid object classes, object names, and attribute names for each instruction.

- For the GSV instruction, you can get values for all the attributes.
- For the SSV instruction, the software displays only those attributes you can set (SSV).

In the S2 file, you could find the data table addresses of system information such as the date and time, forcing status, faults, processor status, I/O status, etc. We know that on a new ControlLogix project there are no tags whatsoever. If you need any status from the system, you will use the GSV command. The GSV command is a simple copy command. The GSV gets data from the system (where we cannot see it). Then it copies the data to a tag where we can see the data and use it in our project.

Let's understand this concept with an industry-based example, I want to know the quality of products being produced in each shift, each shift is of 8 hours.

Example- Shift-wise production report

I want to display the number of products manufactured in first, second, and third shift. So, for that I need the timing of each shift, the time is as follows;

First shift- 6:30 to 14:30 (06:30 am to 02:30 pm)
Second shift- 14:30 to 20:30 (02:30 pm to 10:30 pm)
Third shift- 20:30 to 6:30 (10:30 pm to 06:30 am)

Create tags

- Class name- WallClockTime
- Attribute Name- LocalDateTime
- Destination- Real Clock

I have created this tag in controller, because I want to fetch the data from the controller. This tag will store 7 DINT's (Double Integers), so we will make this tag an array. The seven elements of this tag will contain: Year, Month, Day, Hour, Minute, Second, and Milli Seconds.

Renaming of each double integer. As you can see the controller will store all the data from double integer 0 to 6.

In this case, we'll go to the controller tag database, and in "Edit Tags" Mode, create a tag called Real_clock. You can name the tag anything you like, though, as long as you follow the guidelines for creating a tag. I followed the Same method while creating a SSV instruction, it will let a person to set the clock of controller.

When the controller goes online, it will fetch the timer value from the controller.

As the set time instruction gets true, the system value gets set as you enter using SSV instructions. You can set date as well as time of the controller.

As you can see in the below rung, I have used a set of two GEQ instructions to create a time block for three shifts. When the values of real time clock lie within the limits, the shift time sets accordingly.

Whenever the finished product bit gets true, the counter of the particular shift counts the number of finished products. These production values are stored in the INT value using MOV instructions.

5. Math instructions

- Studio 5000 includes the basic math instructions like Add, Sub, Mul, Div, Square root, Modulo, and compute. Each of these math instructions, except compute instruction, uses two variables.

- The data types you can use in math instructions are SINT, INT, DINT, Floats, and REALs.

a. ADD Instruction

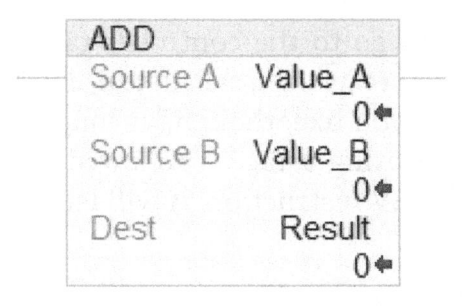

```
   ADD
   Source A    Value_A
                     0←
   Source B    Value_B
                     0←
   Dest        Result
                     0←
```

When enabled, it adds source A to Source B and stores the result in destination.

b. SUB Instruction

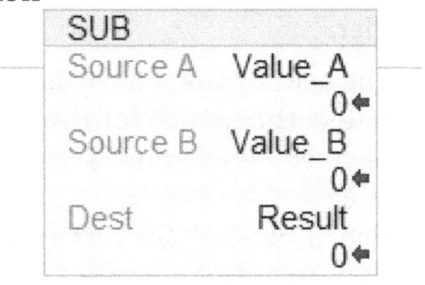

```
   SUB
   Source A    Value_A
                     0←
   Source B    Value_B
                     0←
   Dest        Result
                     0←
```

When enabled, it subtract Source B from Source A and stores the result in destination.

c. MUL Instruction

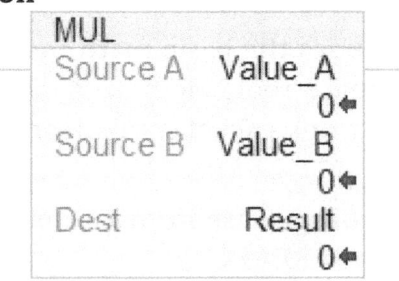

```
   MUL
   Source A    Value_A
                     0←
   Source B    Value_B
                     0←
   Dest        Result
                     0←
```

When enabled, it multiplies Source A with Source B and stores the result in destination.

d. DIV Instruction

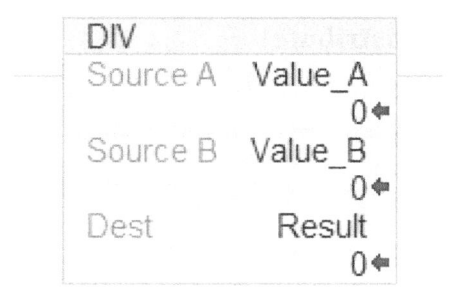

```
DIV
Source A    Value_A
                 0←
Source B    Value_B
                 0←
Dest        Result
                 0←
```

When enabled, it divides Source A by Source B and stores the result in destination.

e. CPT instruction

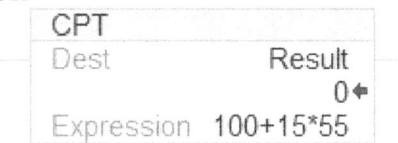

```
CPT
Dest              Result
                      0←
Expression  100+15*55
```

- The CPT (Compute) instruction is useful for condensing equations that would normally take several different math instructions to compose. You define operation in the expression and the result is written in destination.

- For example, to solve this equation (125+50)*(50/2) you may need 3 math instructions. But it can be solved by using just one instruction, that is CPT. You have to simply write this equation in the expression box, and you will get result of this equation once the rung is true.

6. Move instructions

a. MOV (Move)

```
MOV
Source      Value_A
                 0←
Dest        Value_B
                 0←
```

- The move instruction is used to move a value from source to the destination.
- The data type of source and destination should be same.

b. BTD (Bit Field Distribute)

The BTD instruction copies the specified bits from the Source, shifts the bits to the appropriate position, and writes the bits into the Destination.

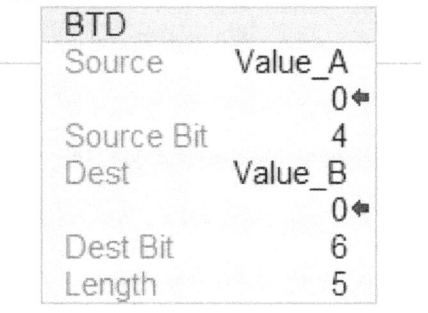

BTD	
Source	Value_A
	0←
Source Bit	4
Dest	Value_B
	0←
Dest Bit	6
Length	5

- Source (DINT/SINT/INT) - Tag that contains bets to move

- Source Bit (DINT) - The number of bit from where to start move.

- Destination (DINT/SINT/INT) - Tag where to move the bits.

- Destination Bit (DINT) - The number of the bit to which the data should be moved must be within the valid range for the Destination data type.

- Length (DINT) - The number of bits to move

Let us understand this instruction with an example. I am using a BTD instruction to move 5 bits from the value 1 to the value 2. This instruction will start moving values from 4 bit of value 1 upto length of 5 bits to the location of value 2.

LEARN EVERYTHING ABOUT PLC PROGRAMMING

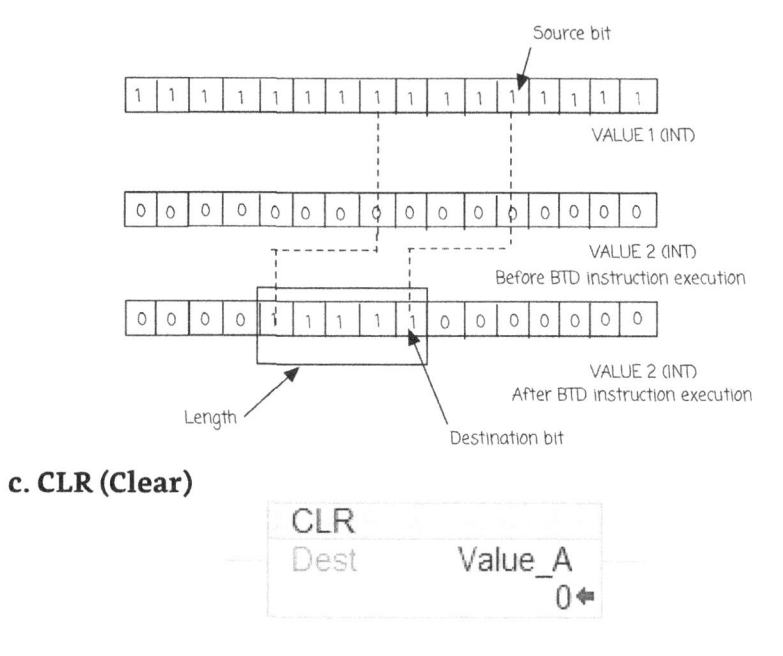

[Diagram showing BTD instruction execution with Source bit, VALUE 1 (INT), VALUE 2 (INT) Before BTD instruction execution, VALUE 2 (INT) After BTD instruction execution, Length, and Destination bit labels]

c. CLR (Clear)

```
CLR
Dest      Value_A
              0
```

This instruction is used to clear the specified bits.

7. Logical instructions

You may be familiar with logic gates which you might have studied before. The studio 5000 logix designer has different logic instructions as shown below.

a. AND

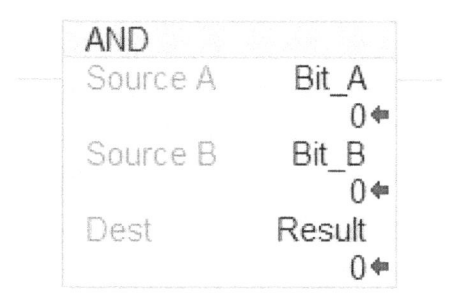

```
AND
Source A    Bit_A
               0
Source B    Bit_B
               0
Dest        Result
               0
```

This instruction gets enabled when both source A and source B bits are available.

b. OR

111

```
OR
Source A    Bit_A
              0
Source B    Bit_B
              0
Dest        Result
              0
```

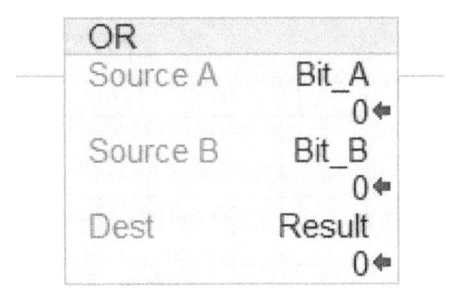

This instruction gets enabled when one of the sources or both sources are available.

c. XOR

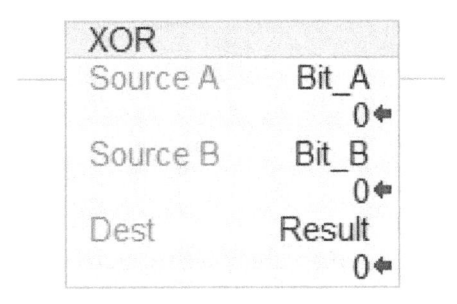

```
XOR
Source A    Bit_A
              0
Source B    Bit_B
              0
Dest        Result
              0
```

This instruction works exactly opposite to OR gate.

d. NOT

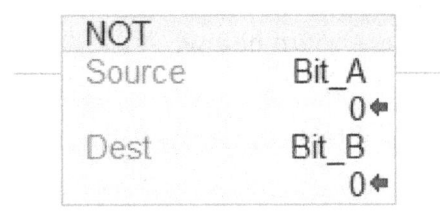

```
NOT
Source      Bit_A
              0
Dest        Bit_B
              0
```

This instruction gets enabled when the source bit is false.

8. File/Shift instructions

a. BSL (Bit Shift Left) and BSR (Bit Shift Right)

The BSL and BSR instructions are used to shift a bit to particular position. A BSL shifts a bit to left position, whereas BSR shifts a bit to the right position within the array.

The source bit's value is loaded into the array's lowest bit by this instruction and the unload bit is determined by the length.

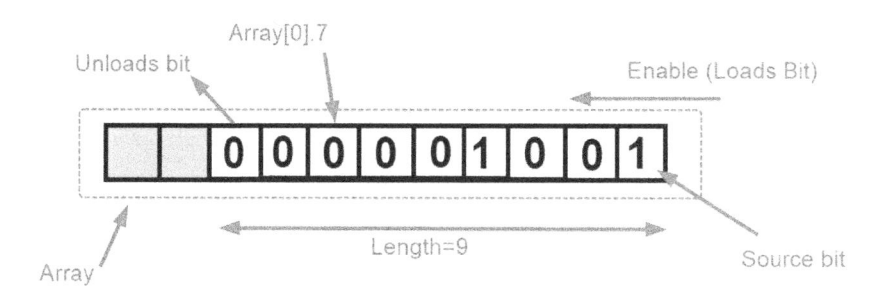

```
         Array[0].7
Unloads bit  |                    Enable (Loads Bit)
   ┌──┬─┬─┬─┬─┬─┬─┬─┬─┐
   │  │0│0│0│0│0│1│0│0│1│
   └──┴─┴─┴─┴─┴─┴─┴─┴─┘
  Array        Length=9          Source bit
```

Box sorting application

Let's understand this concept with real world example. A box sorting system consists of two pneumatic sorters (2 and 3) with a vision system (1). The vision system detects the size of the boxes and conveys it to the controller. The sorter (2) rejects the small boxes whereas sorter (3) rejects the bigger one.

This application looks very simple until a vision system is placed at some distance. There is no any sensor at sorter that will detect the presence of the box. Hence, bit shift instruction plays a vital role here.

Program

Rung 2: I have used a BSL instruction and chosen a length of 9 bits in an array. So, whenever a vision system detects a box, a bit moves to the left by one position. If the size of the box is big then vision system will put value 1 in the first position of the position.

Rung 3: Box sorting
I have kept an eye on third position of the array (BitShiftArray[0].3). When there is value 1 in the third position of the length, the bit gets true and pusher rejects the box.

9. Compare instructions

The compare instructions are used to compare two numbers or strings. One of the values should be a tag value. E.g. comparison between INT and 10.

The comparison instructions include:
a. EQU (Equal)

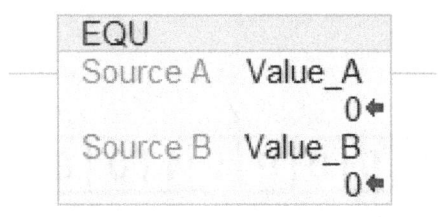

```
EQU
Source A    Value_A
                 0
Source B    Value_B
                 0
```

This instruction enables the output when both values are equal.

b. NEQ (Not Equal)

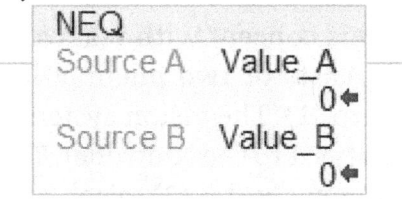

```
NEQ
Source A    Value_A
                 0
Source B    Value_B
                 0
```

This instruction enables the output when both values are not equal.

c. LES (Less than)

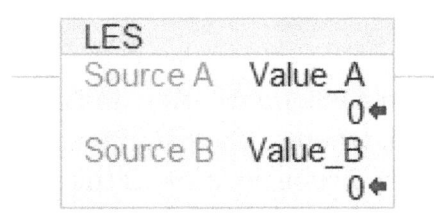

```
LES
Source A    Value_A
                 0
Source B    Value_B
                 0
```

This instruction enables the output when value A is less than value B.

d. GRT (Greater than)

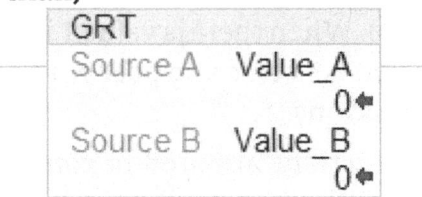

```
GRT
Source A    Value_A
                 0
Source B    Value_B
                 0
```

This instruction enables the output when value A is greater

than value B.

e. LEQ (Less than or equal)

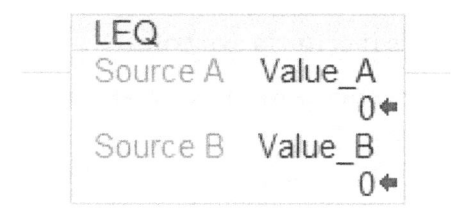

```
LEQ
Source A    Value_A
                 0
Source B    Value_B
                 0
```

This instruction enables the output when value A is less than or equal to value B.

f. GEQ (Greater than or equal)

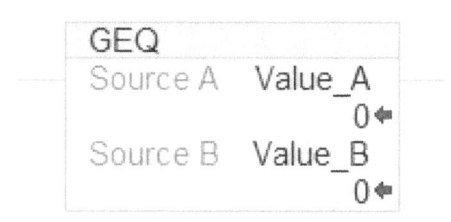

```
GEQ
Source A    Value_A
                 0
Source B    Value_B
                 0
```

This instruction enables the output when value A is greater than or equal to value B.

g. LIM (Limit Test)

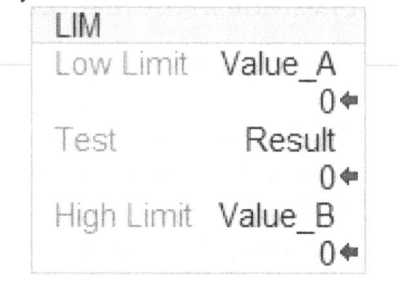

```
LIM
Low Limit     Value_A
                   0
Test          Result
                   0
High Limit    Value_B
                   0
```

The LIM instruction is used to check value within the given tolerance. This instruction enables output when the test value lies within the set limits.

10. Program control instructions

In this section we are going to cover program control instructions such as Jump (JMP), Label (LBL), Jump to subroutine (JSR), Subroutine (SBR), and Return (RET). These instructions are need to be set up in pairs in order to work. These instructions are used to skip segments of code, and only run those segments if criteria are true.

a. Jump (JMP) and Label (LBL) instructions

These two instructions are used as a pair in a routine. As you can see in the below example, when the jump instruction is initiated, the program will be executed from rung number 3 and second rung will be skipped.

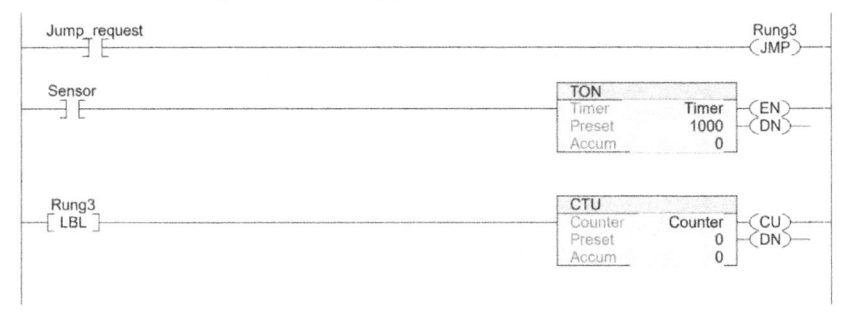

```
Jump_request                                                    Rung3
    ] [                                                          (JMP)

Sensor                                          TON
    ] [                                         Timer      Timer  (EN)
                                                Preset     1000   (DN)
                                                Accum         0

Rung3                                           CTU
   [ LBL ]                                      Counter   Counter (CU)
                                                Preset        0   (DN)
                                                Accum         0
```

b. JSR (Jump to subroutine) and SBR (Subroutine)

To use JSR instruction you must pair it with SBR instruction. And whichever routine you are jumping to must have SBR instruction the first position in the first rung.

▲ 📁 Tasks
    ▲ ⚙ MainTask
        ▲ 📖 MainProgram
            ◇ Parameters and Local Tags
            ▣ MainRoutine
            ▤ A_Auto_Mode
            ▤ B_Man_Mode
            ▤ C_Production_Report
            ▤ D_Alarms

As you can see here I have created separate routine for different programming steps.

Now I have to call all the routines in main program in order to run the program inside them using JSR instruction.

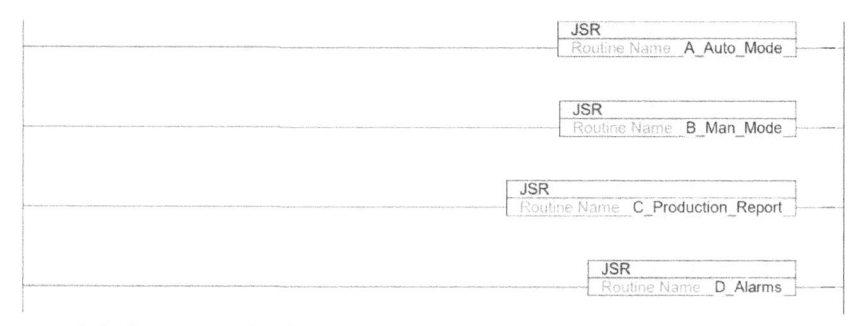

```
                                    | JSR                              |
                                    | Routine Name  A_Auto_Mode        |

                                    | JSR                              |
                                    | Routine Name  B_Man_Mode         |

                            | JSR                                      |
                            | Routine Name  C_Production_Report        |

                                    | JSR                              |
                                    | Routine Name  D_Alarms           |
```

c. AFl (Always False)

You can use AFl instruction to temporarily disable a rung while you are debugging a program. This instruction disables the all the instructions on this rung.

```
           Sensor                              | TON                    |
   -[ AFI ]--] [-----------------------------  | Timer       Timer -(EN)|
                                               | Preset          0 -(DN)|
                                               | Accum           0      |

   Timer.DN                                                        Lamp
   --] [-----------------------------------------------------------( )--
```

d. MCR (Master Control Reset)

The MCR instruction is like a reset bit, but this instruction resets multiple coil at a time which are within the MCR zone. You have to use two MCR instructions to create a zone.

As you can see I have create a MCR zone using two instructions when inputs 1 and 2 are false, the rungs will operate with normal condition but when both inputs are true, the outputs will be cleared regardless of input conditions within that zone.

```
Input_1   Input_2
──┤ ├──────┤ ├──────────────────────────────────(MCR)──

  Input_3                          ┌─TON─────────────┐
──┤ ├────────────────────────────  │ Timer    Timer  │─(EN)─
                                   │ Preset      0   │─(DN)─
                                   │ Accum       0   │
                                   └─────────────────┘
                                   Timer.DN              Out_1
                                  ──┤ ├──────────────────( )──

  Input_4                                                Out_2
──┤ ├───────────────────────────────────────────────────(L)──

  Input_5                                                Out_2
──┤ ├───────────────────────────────────────────────────(U)──

─────────────────────────────────────────────────────────(MCR)──
```

5.1 Types of Data

When you click on Assets in the controller organizer, you will see different data types. We are going to learn a few data types that you must know about.

- Module defined
- Pre-defined
- User defined

1. Module defined

The module defined tags get generated with the configuration of the module. This module data can be seen in the controller tag screen, also it get assigned with the modules which you have used in your project. This data type can not be changed.

2. Pre-defined

The predefined tags are already present in the folder when you open a software. These tag includes counter, timer, alarm, etc. You can not modify these tags.

- Predefined
    - ALARM
    - ALARM_ANALOG
    - ALARM_DIGITAL
    - ALARM_SET_CONTROL
    - AUX_VALVE_CONTROL
    - AXIS_CIP_DRIVE
    - AXIS_CONSUMED
    - AXIS_GENERIC
    - AXIS_GENERIC_DRIVE
    - AXIS_SERVO
    - AXIS_SERVO_DRIVE
    - AXIS_VIRTUAL
    - BOOL
    - BUS_OBJ
    - CAM
    - CAMSHAFT_MONITOR
    - CAM_PROFILE
    - CB_CONTINUOUS_MODE
    - CB_CRANKSHAFT_POS_MONITOR
    - CB_INCH_MODE
    - CB_SINGLE_STROKE_MODE
    - CC
    - CONFIGURABLE_ROUT
    - CONNECTION_STATUS
    - CONTROL
    - COORDINATE_SYSTEM
    - COUNTER

3. User defined

We are already familiar with the basic types of data such as bool, SINT, INT, etc. A User defined Data (UDT) is user-created data type that contains some or all these basic data types. It is like creating one big data type that contains smaller data types within it. This comes in handy when you have to repeat same types of structure within a program multiple time. It helps a programmer to save time.

Let's understand this concept with an example.

Example:

A machine consists of Red and Green lamps. Both lamps follow the same flow of operating sequence using two timers. So, instead of creating separate data for timers and bools, let's create only onw data type.

A. Create a UDT

1. Right click on the user defined option and select 'New Data type'.

2. Name the data type as 'Lamp'.

3. Create a list of members inside Lamp data.

2. Select the data type, such as Bool and timer.

3. Click apply and OK.

Name	Data Type	Description
ON	BOOL	
OFF	BOOL	
On_Delay	TIMER	
Off_Delay	TIMER	
Add Member...		

B. Configure tags in controller

LEARN EVERYTHING ABOUT PLC PROGRAMMING

Name	Alias For	Base Tag	Data Type	Class
Conveyor			BOOL	Standard
▷ Conveyor:I			AB:PowerFlex525...	Standard
▷ Conveyor:O			AB:PowerFlex525...	Standard
▷ Drive:I			AB:PowerFlex525...	Standard
▷ Drive:O			AB:PowerFlex525...	Standard
▲ Red_Lamp			Lamp	Standard
Red_Lamp.ON			BOOL	Standard
Red_Lamp.OFF			BOOL	Standard
▷ Red_Lamp.On_Delay			TIMER	Standard
▷ Red_Lamp.Off_Delay			TIMER	Standard
▲ Green_Lamp			Lamp	Standard
Green_Lamp.ON			BOOL	Standard
Green_Lamp.OFF			BOOL	Standard
▷ Green_Lamp.On_Delay			TIMER	Standard
▷ Green_Lamp.Off_Delay			TIMER	Standard

While creating a tag in controller IO, just name a tag like red lamp and green lamp. Select a data type as lamp. As you can see here, all the data member tags get attached with the controller defined tags.

For example- **Green_Lamp.On_delay**.

C. Program using UDT

After creating an user defined tag, it has helped me to write a program with less efforts. This data type is very useful when you write a repeated program structure.

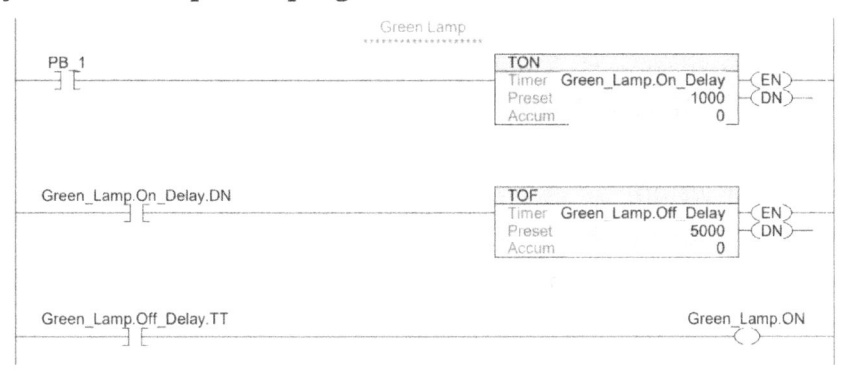

```
                           Green Lamp
   PB_1                                    TON
 ──┤ ├──────────────────────────────────  Timer   Green_Lamp.On_Delay ─(EN)─
                                           Preset              1000   ─(DN)─
                                           Accum                 0

  Green_Lamp.On_Delay.DN                   TOF
 ──┤ ├──────────────────────────────────  Timer   Green_Lamp.Off_Delay ─(EN)─
                                           Preset              5000   ─(DN)─
                                           Accum                 0

  Green_Lamp.Off_Delay.TT                                  Green_Lamp.ON
 ──┤ ├──────────────────────────────────────────────────────────( )──
```

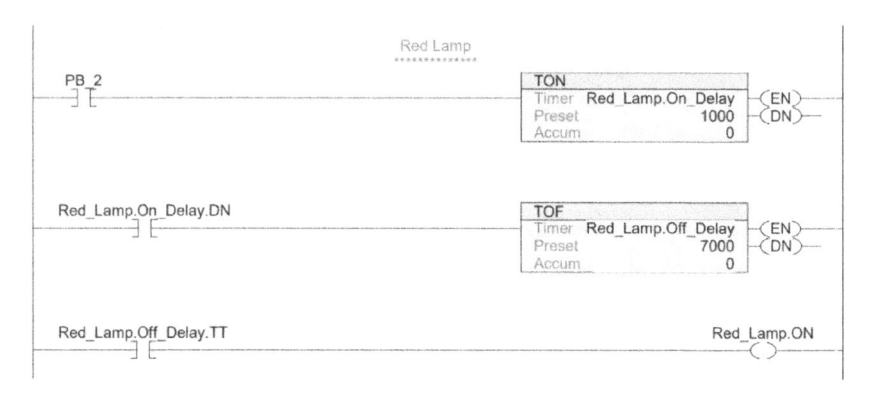

```
                              Red Lamp
  PB_2                                    ┌─TON─────────────────────┐
──┤ ├──────────────────────────────────────┤ Timer  Red_Lamp.On_Delay├─(EN)──
                                           │ Preset            1000  ├─(DN)──
                                           │ Accum                0  │
                                           └─────────────────────────┘

  Red_Lamp.On_Delay.DN                    ┌─TOF──────────────────────┐
──┤ ├──────────────────────────────────────┤ Timer  Red_Lamp.Off_Delay├─(EN)──
                                           │ Preset            7000  ├─(DN)──
                                           │ Accum                0  │
                                           └──────────────────────────┘

  Red_Lamp.Off_Delay.TT                                        Red_Lamp.ON
──┤ ├──────────────────────────────────────────────────────────────( )──
```

5.2 Add on instructions (AOI)

The Add on instruction is user created instruction. It is like a mini program that has its own local tags and routine. AOI is used when you need a simple system, multiple times.

The most powerful features of the studio 5000 Logix designer program is the ability to create Add-On Instructions. This allows a programmer to define an instruction that contains a commonly used function or algorithm (as a set of instructions), and use it as one instruction.

Let's understand the AOI concept with a very simple example. A system consists of a conveyor with 3 push buttons (Fwd, rev, and stop). When an operator has pressed start button, the conveyor will start after a few time period for specific time. Also the conveyor will move reverse when the reverse PB is pressed.

In the both operation we need two timers (On delay and off delay). If you look closely, the operation has repeated sequence of operation. These repeated sequence can be avoided using AOI instruction.

Now we are going to see the AOI operation in steps:

1. Create Add on instructions

To create an AOI, make sure you have the controller organizer window open and go down to the Add on instruction folder. Right click on this folder and you will see the following pop-up appear.

LEARN EVERYTHING ABOUT PLC PROGRAMMING

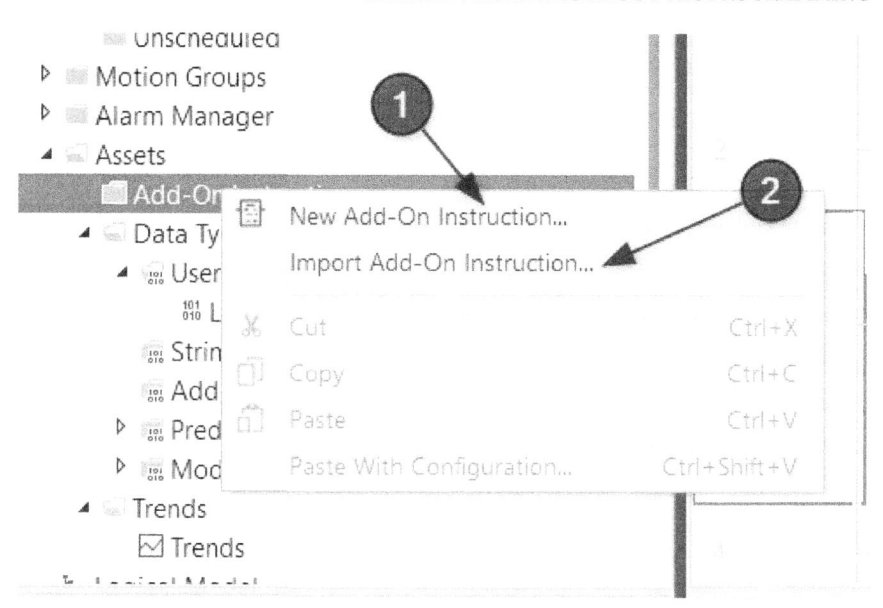

After selecting 'New Add-on instruction' a new window will appear, where you have to give name to the AOI, select program type, and click OK.

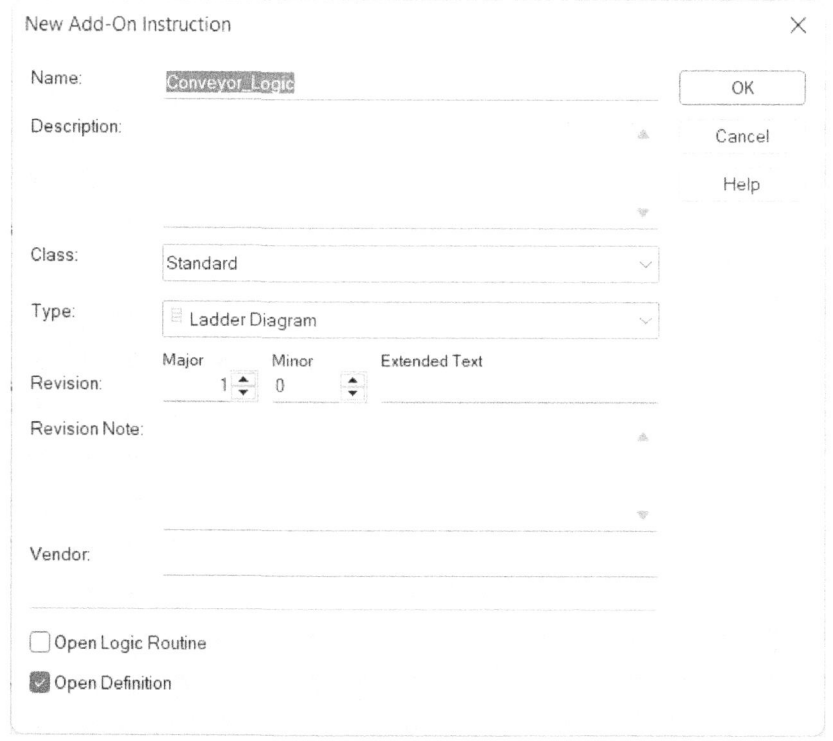

123

2. Write logic in AOI folder

When the conveyor start bit is true, the value will be moved from 'Conv_On_Delay' to the Timer 1, and the Timer1.DN bit will get true after that set timer.

When the Timer1.DN bit is true, the value will be moved from 'Conv_Off_Delay' to the Timer 2, and the 'Conv_ON' bit will be true for that time period.

3. AOI Parameters

In order to run the above logic we have to set the parameters of the AOI. The parameters are as follows:

Conv_Start (Bool) - Input paramter
Conv_on_Delay (DINT) - Timer1 value
Conv_off_Delay (DINT) - Timer2 value
Conv_ON (Bool) - Output parameter

Select the parameters you want to show in AOI box by selecting the parameters as shown in the image below.

4. Create an UDT

In order to work this logic we need to create 2 UDTs in the controller tag.

1. Conveyor_FWD
2. Conveyor_REV

Once you create an UDT with conveyor as data type, we will see the data as follows:

5. Deployment of AOI in ladder logic

Let's understand the working of a conveyor after adding AOI in our logic. I have used signle AOI instruction to run a conveyor in forward and reverse direction.

1. Forward direction

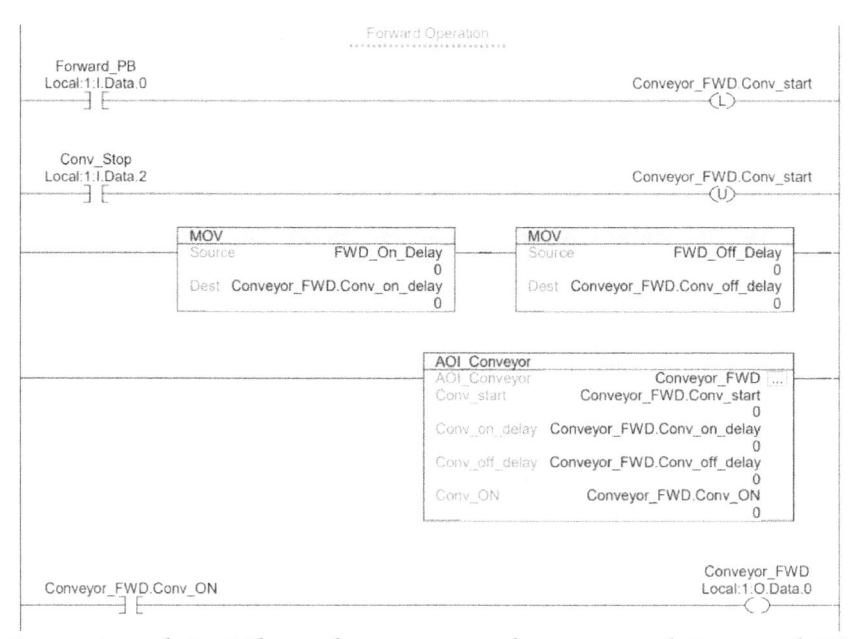

```
                                    Forward Operation
  Forward_PB
  Local:1:I.Data.0                                          Conveyor_FWD.Conv_start
      ] [                                                              (L)

  Conv_Stop
  Local:1:I.Data.2                                          Conveyor_FWD.Conv_start
      ] [                                                              (U)

                    ┌─ MOV ──────────────────┐    ┌─ MOV ──────────────────┐
                    │ Source       FWD_On_Delay│   │ Source      FWD_Off_Delay│
                    │                        0 │   │                        0 │
                    │ Dest  Conveyor_FWD.Conv_on_delay │ │ Dest  Conveyor_FWD.Conv_off_delay │
                    │                        0 │   │                        0 │
                    └────────────────────────┘    └────────────────────────┘

                              ┌─ AOI_Conveyor ─────────────────────────────┐
                              │ AOI_Conveyor              Conveyor_FWD ...│
                              │ Conv_start       Conveyor_FWD.Conv_start  │
                              │                                        0  │
                              │ Conv_on_delay  Conveyor_FWD.Conv_on_delay │
                              │                                        0  │
                              │ Conv_off_delay Conveyor_FWD.Conv_off_delay│
                              │                                        0  │
                              │ Conv_ON          Conveyor_FWD.Conv_ON     │
                              │                                        0  │
                              └────────────────────────────────────────────┘

                                                          Conveyor_FWD
  Conveyor_FWD.Conv_ON                                    Local:1:O.Data.0
         ] [                                                     ( )
```

Rung 1 and 2- When the operator has pressed Forward PB, the 'conveyor_FWD.Conv_start' bool gets latched, also it gets unlatched when the 'Conv_stop' PB is pressed.

Rung 3- At the same time 'FWD_On_Delay' and 'FWD_Off_Delay' timers values will be moved into the timers 1 and 2 of an AOI.

Rung 4 and 5- Once the 'conveyor_FWD.Conv_start' is true, the AOI will start executing the logic inside the folder and start running a conveyor in forward direction.

2. Reverse direction

Rung 6 and 7- When the operator has pressed Reverse PB, the 'conveyor_REV.Conv_start' bool gets latched, also it gets unlatched when the 'Conv_stop' PB is pressed.

Rung 8- At the same time 'REV_On_Delay' and 'REV_Off_Delay' timers values will be moved into the timers 1 and 2 of an AOI.

Rung 9 and 10- Once the 'conveyor_REV.Conv_start' is true, the AOI will start executing the logic inside the folder and start running a conveyor in reverse direction.

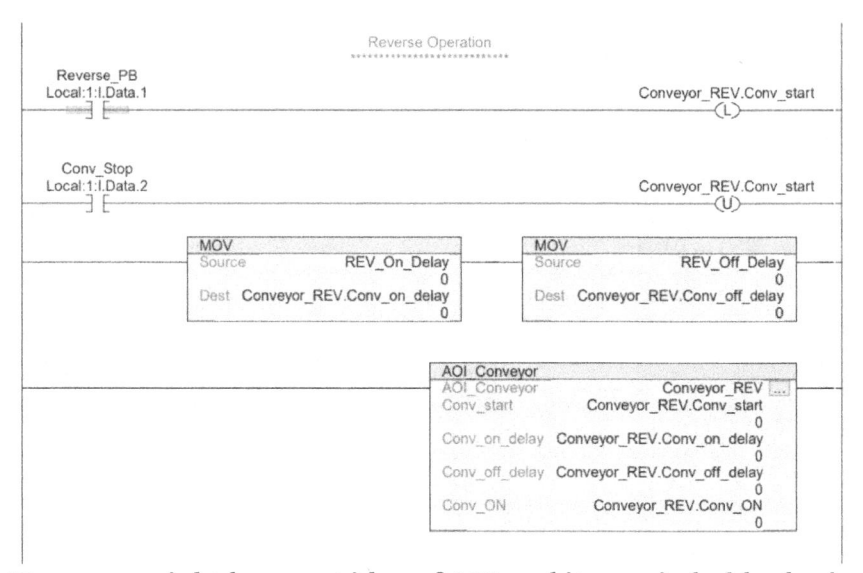

```
                          Reverse Operation
  Reverse_PB
  Local:1:I.Data.1                                      Conveyor_REV.Conv_start
  ──┤ ├──────────────────────────────────────────────────────────(L)──

  Conv_Stop
  Local:1:I.Data.2                                      Conveyor_REV.Conv_start
  ──┤ ├──────────────────────────────────────────────────────────(U)──

                    ┌─MOV──────────────────┐   ┌─MOV──────────────────┐
                    │ Source   REV_On_Delay│   │ Source   REV_Off_Delay│
                    │                    0 │   │                    0 │
                    │ Dest Conveyor_REV.Conv_on_delay│ Dest Conveyor_REV.Conv_off_delay│
                    │                    0 │   │                    0 │
                    └──────────────────────┘   └──────────────────────┘

                              ┌─AOI_Conveyor──────────────────────────┐
                              │ AOI_Conveyor              Conveyor_REV│
                              │ Conv_start    Conveyor_REV.Conv_start │
                              │                                     0 │
                              │ Conv_on_delay Conveyor_REV.Conv_on_delay│
                              │                                     0 │
                              │ Conv_off_delay Conveyor_REV.Conv_off_delay│
                              │                                     0 │
                              │ Conv_ON       Conveyor_REV.Conv_ON    │
                              │                                     0 │
                              └───────────────────────────────────────┘
```

Now, you might have got idea of AOI and its use in ladder logic.

2. Import Add on instructions

Right click on the Add on instruction and select import add on instruction. The add on instruction file has L5X extension. This feature is helpful when you have a repeated program in different project.

Many third-party vendors and OEMs release custom AOIs for their equipment. An example which comes to mind is that Cognex In-Sight Camera AOI. This instruction allows the user to easily interface with the camera supplied by Cognex. This practice is common & provides integrators with a simple way to interface with a certain device without having to work with registers directly.

Source code protection for AOI

In a particular case, if a programmer wants to protect Add on instructions that contained some unique logic, the Rockwell automation provides an easy tool that comes with it software itself. Since protecting the intellectual property or touchy algorithms while opening up some of the code for modification is pretty common. I'll walk you through configuring an AOI for source protection, you can use the same approach to protect routines as well.

To check if AOI is source protected

1. Select AOI in the controller organizer.
2. Look in the quick view panel for protection type.

Protection type field indicates if the AOI is protected by a license or a source key. If the protection attribute is not listed, then the instruction is not protected.

To protect your AOI, right click on the AOI instruction and select properties.

When you click on the signature tab, the another screen will appear where you have click of signature to protect your application.

J. Powerflex 525 VFD Programming using studio 5000

I'll demonstrate how to use the CompactLogix PLC and ethernet communication to control the Powerflex 525 VFD in this section.

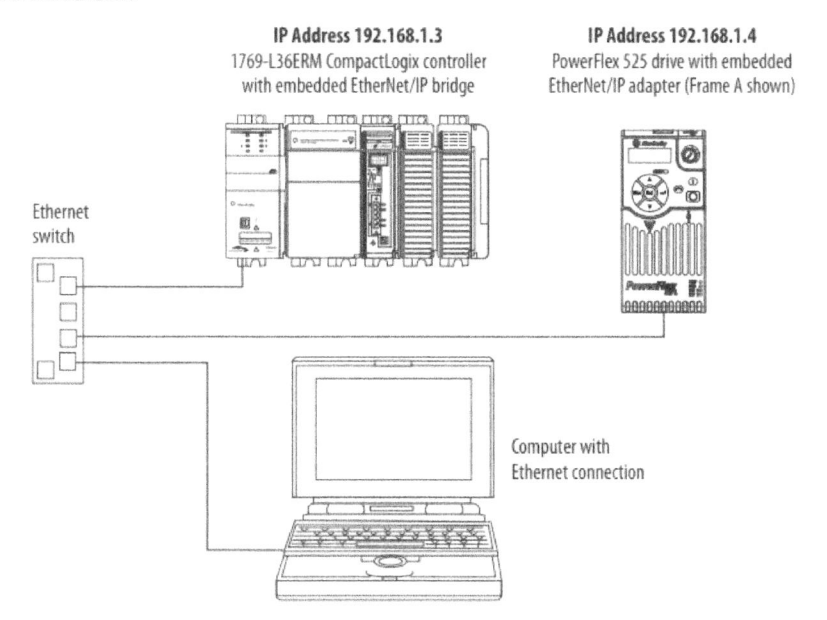

Step 1 : Set IP address of VFD using BootP tool or Manually

The initial step of configuration is to set the VFD's IP address when it is turned on. You can follow two methods to set an IP address of the VFD.

a. BootP tool
1. Connect Ethernet cable to your computer and the VFD.
2. The default parameter C128 should be 2 (BootP).
3. Launch BootP tool to detect the MAC address of the VFD and then change the IP address.
4. Give a power cycle to save the IP address.

b. Set IP address from VFD
The communication parameter is denoted by letter C. You must set these communication parameters from the VFD screen.

Parameter	Setting	Description
C128	01	Verify that parameter C128 is set to 1. This parameter must be set to "Parameters" to configure the IP address using parameters.
C129	192	
C130	168	
C131	01	IP Address
C132	62	
C133	255	
C134	255	
C135	255	Subnet Mask
C136	0	
C137	192	
C138	168	
C139	01	Gateway Address
C140	01	

Note: Set communication parameter C128 to 2, so that when the VFD is powered on again it will retrive communication paramters from the VFD itself.

Once you are done, restart the device to save the configuration. Now, you can see Link LED is solid and ENET LED is flashing on the VFD.

Step 2 : Change IP address mode using RS Linx classic

Once you perform a power cycle, the VFD will lose its IP address, so you must switch the IP mode using the RS Linx program.
a. Launch the classic version of RS Linx and set the ethernet drivers. b. The connected VFD can be seen in the RS Linx classic software.
c. Module configuration can be found by doing a right-click on the VFD. Change the IP address's dynamic to static mode.

Step 3 : Set Operation parameters

You can set paramters of VFD using 3 ways:

- From VFD
- Using CCW software
- Using Studio 5000 Logix designer

Parameters that you need to set:

Parameter	Description	Value
P30	Language There are 15 languages	English is default language
P41	Acceleration time Sets the time for the drive to accel from 0 Hz to [Maximum Freq]	The time is in seconds
P42	Deceleration time Sets the time for the drive to accel from [Maximum Freq] Hz to 0	The time is in seconds
P43	Minimum frequency	It should not be less than 0.0 Hz
P44	Maximum frequency	It should not be less than 60 Hz
P46	Start source	1 = Keypad 2 = DigIn TrmBlk 3 = Serial/DSI 4 = Network Opt 5 = Ethernet/IP
P47	Speed reference	1 = Drive Pot 2 = Keypad Freq 3 = Serial/DSI 4 = Network Opt 5 = 0-10V Input 6 = 4-20mA Input 7 = Preset Freq 8 = Anlg In Mult 9 = MOP 10 = Pulse Input 11 = PID1 Output 12 = PID2 Output 13 = Step Logic 14 = Encoder 15 = Ethernet/IP 16 = Positioning

Step 4 : Configure powerflex VFD in studio 5000 software

a. Open studio 5000 Logix Designer. Right click on the Ethernet and select the VFD from the list
b. Enter the IP address and give it a name.
c. Click OK.
d. Controller tags - Once you complete the configuration, you can see the Input and Output tags for the powerflex 525 VFD. You can control and monitor a VFD using these tags.

Step 5 : PLC program

I have developed a HMI screen, that offers:
- Monitoring the status of VFD
- Controlling the speed of the VFD
- Controlling the direction of the Motors though

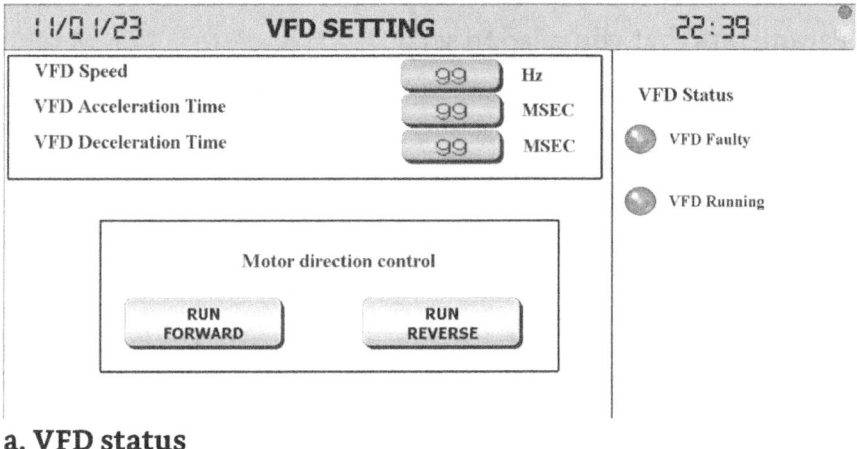

a. VFD status

I have used VFD status to set the drive ready and used this tag to display its status on HMI.

b. VFD direction control

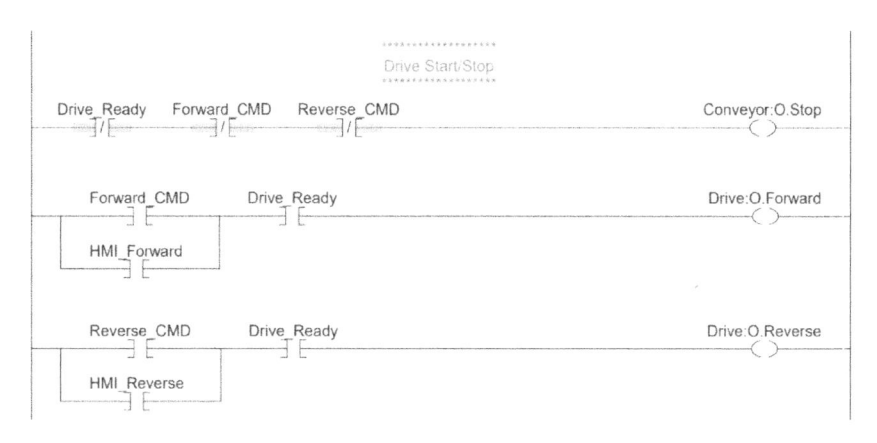

```
                              Drive Start/Stop
   Drive_Ready   Forward_CMD   Reverse_CMD                       Conveyor:O.Stop
      ─┤/├──────────┤/├──────────┤/├──────────────────────────────────( )──

        Forward_CMD      Drive_Ready                              Drive:O.Forward
      ────┤ ├────────────┤ ├──────────────────────────────────────────( )──
        HMI_Forward
      ────┤ ├──

        Reverse_CMD      Drive_Ready                              Drive:O.Reverse
      ────┤ ├────────────┤ ├──────────────────────────────────────────( )──
        HMI_Reverse
      ────┤ ├──
```

Rung 1- The VFD will remain in stop mode, if any of three conditions is true.

Rung 2- If the drive is ready and one of the bits of forward CMD and HMI forward is true, then VFD will start moving forward.

Rung 3- If the drive is ready and one of the bits of reverse CMD and HMI reverse is true, then VFD will start moving forward.

c. Speed control

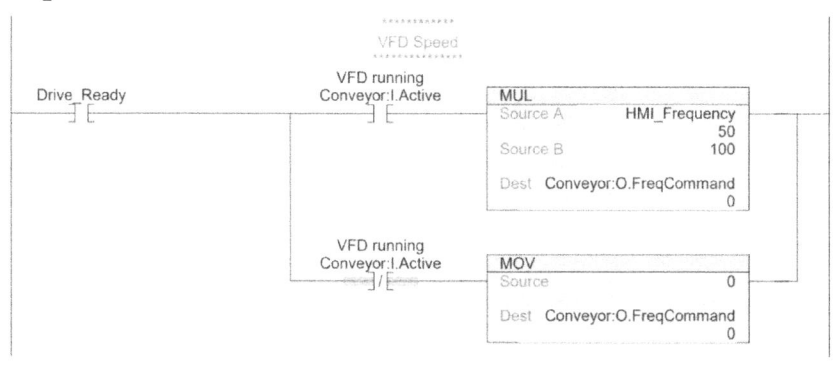

```
                              VFD Speed
                        VFD running
   Drive_Ready          Conveyor:I.Active      ┌─ MUL ──────────────────────┐
   ────┤ ├──────────────────┤ ├────────────────│ Source A     HMI_Frequency │
                                               │                         50 │
                                               │ Source B               100 │
                                               │ Dest  Conveyor:O.FreqCommand│
                                               │                          0 │
                                               └────────────────────────────┘
                        VFD running
                        Conveyor:I.Active      ┌─ MOV ──────────────────────┐
                      ──────┤/├────────────────│ Source                   0 │
                                               │ Dest  Conveyor:O.FreqCommand│
                                               │                          0 │
                                               └────────────────────────────┘
```

Rung 1- When the drive is ready and in running condition, the frequency which is entered from HMI screen by operator will be moved to VFD in INT data format after multiplying it by 100.

It is necessary to multiple the value by 100. Similarly, the value 0 will be moved to the VFD if it is not in active state. You have to configure HMI frequency with value enter tab in HMI while developing HMI screens.

CASE STUDY 2

Elevator

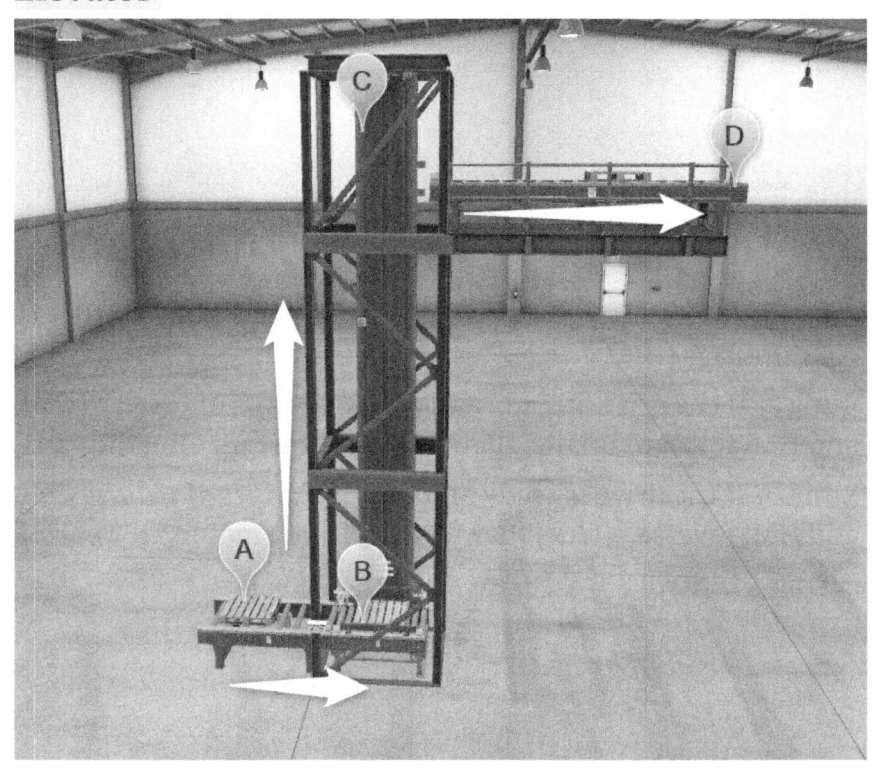

In this case study, I have considered an example of an elevator. The elevator is used to transfer a pallet from ground floor to the top floor. It consists of:

A. Infeed conveyor - This conveyor feeds the boxes to the elevator.

B. Roller conveyor- The carries the boxes.

C. Elevator- It consists of a motor with pulley belt mechanism for up down movement of the roller conveyor.

D. Outfeed Conveyor- It transfers the boxes from roller conveyor to another platform.

Communication Architecture
- PLC - CompactLogix 5380
- Remote IO - 1734 AENTR

- Input cards IB8 and Output card OB8
- HMI- PanelView
- VFD - Powerflex 525

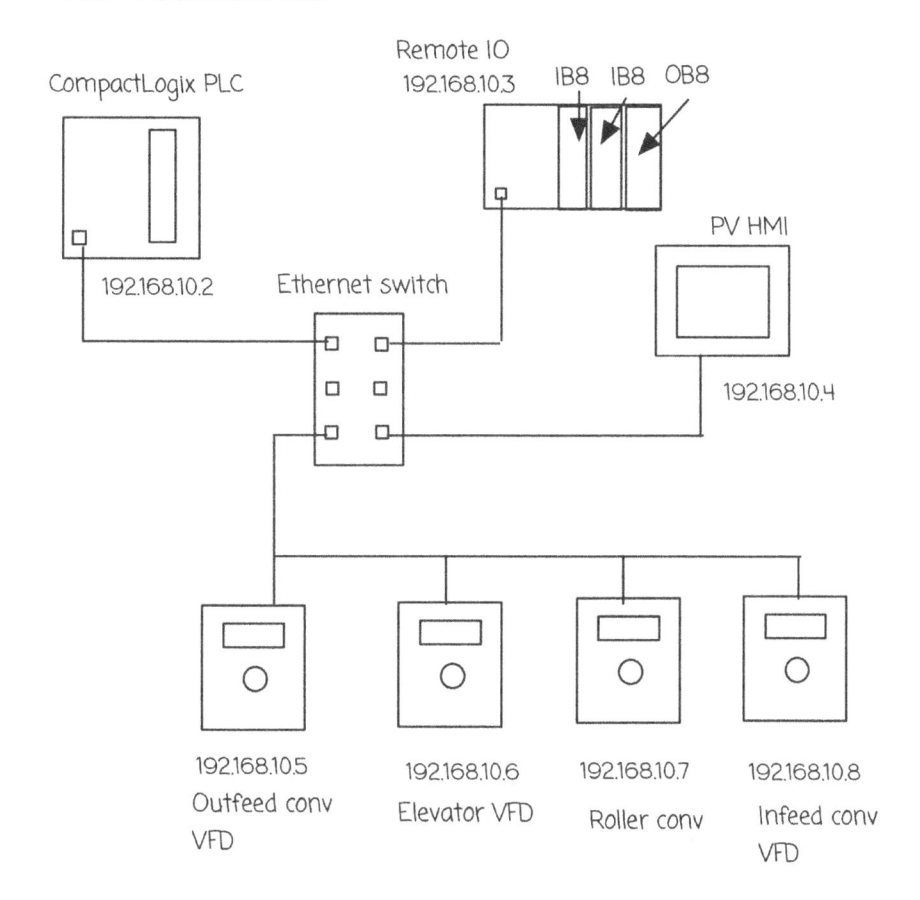

```
CompactLogix PLC        Remote IO
                        192.168.10.3    IB8  IB8  OB8

192.168.10.2   Ethernet switch                 PV HMI

                                               192.168.10.4

192.168.10.5      192.168.10.6    192.168.10.7    192.168.10.8
Outfeed conv      Elevator VFD    Roller conv     Infeed conv
VFD                                               VFD
```

Operation

- If the roller conveyor is at the bottom (Bottom sensor1 is true) and the product present sensor is false, the infeed conveyor will transfer a pallet to the roller conveyor when the elevator is in automatic mode.
- The pallet will begin to rise as it reaches the product presence sensor. As it approaches high sensor 1, the lifter slows down and it stops completely When the pallet reaches high sensor 2.
- The roller conveyor will begin unloading the pallet after the high sensor 2 has stabilized. As soon as the pallet is unloaded,

the roller conveyor will halt and begin to descend. As it approaches Bottom sensor 2, it will slow down and eventually come to a complete halt. Likewise the cycle repeats in auto mode.
- A selector switch can be put in manual mode to move the lifter in upward and downward direction.

Configuration of devices

In this section, we are going to see the configuration of different peripheral devices such as ethernet adapter, IO cards, and VFDs.

a. PLC controller
When you open the studio 5000 Logix designer software, this screen will appear in front of you.
1. Simple enter the controller part number in the search box, and select the controller.
2. Name the project and click next button.

b. Remote IO configuration
1. Right click on the A1 ethernet and select add module, give a name to the adapter.

2. Set IP address, and click OK.

3. IO cards configuration
We have total 5 IO cards to configure. You can refer chapter no. 3 to learn about configuration of IO cards.

<u>Change chassis size</u>
1. Right click on the Ethernet Adapter
2. As you can see the current chassis size is 1. Click on the 'change' button to increase the size of chassis.
3. Now select the chassis size 6 (AENT + 5 IO cards) and click OK.

Add IO cards
Now the next step is to fit the IP cards into the 5 different slots of Ethernet adapter. Follow the procedure to add the IO cards.

1. Right click on the chassis and select add module.
2. In the search box, type the module description, and click on

create tab.
4. VFD configuration
a. Set IP address of VFD using BootP
You can set IP address of the VFD from keypad or using BootP tool.
b. Add VFD in project
Open the project in studio 5000 and perform following steps to add VFDs in your project.
a. Right click on the Ethernet and select add module.
b. Select the VFD from the list and enter details of the VFD with its IP address.
c. Download the Program in PLC and check if the VFD is in communication network with PLC.
C. VFD Parameters
Apart from the motor's parameters we have to change following parameters to establish the communication between PLC and VFD via ethernet communication.
1. P46 (Start source) - 5 (Ethernet)
2. P47 (Speed reference) - 15 (Ethernet)

Ladder Logic

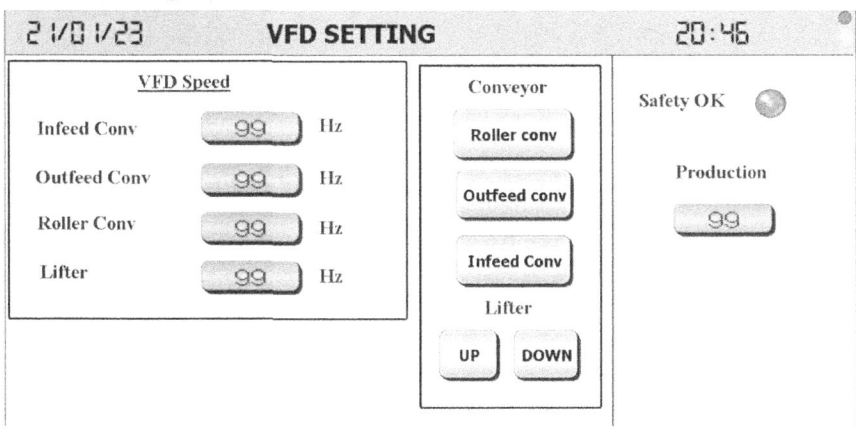

```
2 1/0 1/23            VFD SETTING              20:46

        VFD Speed                Conveyor
   Infeed Conv   [ 99 ] Hz      [ Roller conv ]     Safety OK  ◯
   Outfeed Conv  [ 99 ] Hz                           Production
                                 [ Outfeed conv ]
   Roller Conv   [ 99 ] Hz                            [ 99 ]
   Lifter        [ 99 ] Hz      [ Infeed Conv ]
                                    Lifter
                                 [ UP ] [ DOWN ]
```

The speed of the conveyors and elevator can be changed from HMI screen.

a. IO mapping

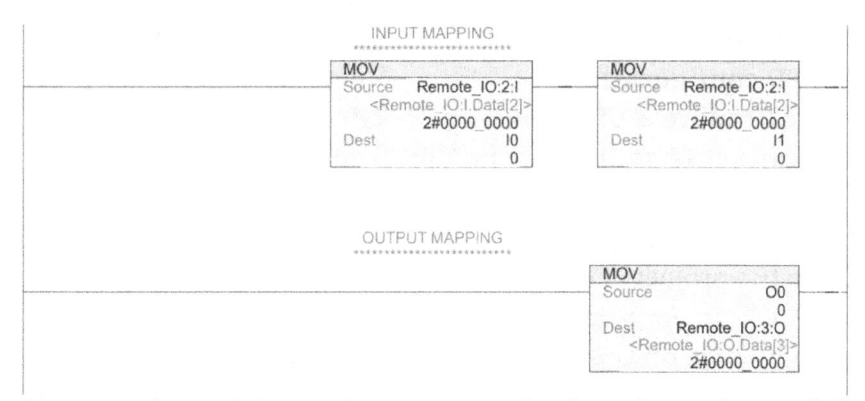

```
                    INPUT MAPPING
           ┌─────────────────────────┐  ┌─────────────────────────┐
           │ MOV                     │  │ MOV                     │
           │ Source   Remote_IO:2:I  │  │ Source   Remote_IO:2:I  │
           │         <Remote_IO:I.Data[2]>│        <Remote_IO:I.Data[2]>│
           │              2#0000_0000│  │              2#0000_0000│
           │ Dest                  I0│  │ Dest                  I1│
           │                        0│  │                        0│
           └─────────────────────────┘  └─────────────────────────┘


                    OUTPUT MAPPING
                                        ┌─────────────────────────┐
                                        │ MOV                     │
                                        │ Source                O0│
                                        │                        0│
                                        │ Dest     Remote_IO:3:O  │
                                        │         <Remote_IO:O.Data[3]>│
                                        │              2#0000_0000│
                                        └─────────────────────────┘
```

I have used Mov instruction to move the data from the module to the SINT data blocks.

Tag	Value	Type	Description
▲ I0	0	Deci... SINT	
I0.0	0	Deci... BOOL	EMG PB
I0.1	0	Deci... BOOL	RESET PB
I0.2	0	Deci... BOOL	AUTO/MAN SS
I0.3	0	Deci... BOOL	START PB
I0.4	0	Deci... BOOL	STOP PB
I0.5	0	Deci... BOOL	TOP LS
I0.6	0	Deci... BOOL	BOTTOM LS
I0.7	0	Deci... BOOL	BOTTOM STOP SENSOR
▲ I1	0	Deci... SINT	
I1.0	0	Deci... BOOL	BOTTOM SLOW SENSOR
I1.1	0	Deci... BOOL	TOP STOP SENSOR
I1.2	0	Deci... BOOL	TOP SLOW SENSOR
I1.3	0	Deci... BOOL	PRODUCT PRESENT SENSOR
I1.4	0	Deci... BOOL	INFEED SENSOR
I1.5	0	Deci... BOOL	OUTFEED SENSOR
I1.6	0	Deci... BOOL	ELEVATOR UP
I1.7	0	Deci... BOOL	ELEVATOR DOWN

To make program write more easy, I have gives names to each and every boolean tags so that I can use them in program directly.

b. Program Routine

The elevator will start working properly when the all the interlocks given below are true.

```
                          OPERATION MODE

  EMG PB    TOP LS   BOTTOM LS
   I0.0      I0.5      I0.6     Elevator:I.Faulted  Infeed_Conv:I.Faulted  Outfeed_conv:I.Faulted
   ─┤ ├──────┤ ├──────┤ ├───────────┤ ├─────────────────┤ ├──────────────────┤ ├─
           Roller_Conv:I.Faulted                                              SAFETY_OK
                 ─┤ ├─                                                         ─(L)─

  RESET PB
   I0.1      ONS.0                                                            SAFETY_OK
   ─┤ ├─────[ONS]─                                                             ─(U)─
```

The elevator operates in manual or auto mode by a selector switch.

```
             AUTO/MAN SS
  SAFETY_OK     I0.2                                                          MAN_MODE
   ─┤ ├─────────┤ ├─                                                           ─( )─

                               AUTO/MAN SS
  SAFETY_OK                       I0.2         ONS.0      AUTO_MODE
   ─┤ ├──────────────────────────┤/├─────────[ONS]─         ─(L)─
                                STOP PB
                                 I0.4                      AUTO_MODE
                                ─┤ ├─                       ─(U)─
```

When the elevator is in auto mode, the infeed and outfeed conveyor will start running on the frequency entered from the HMI screen.

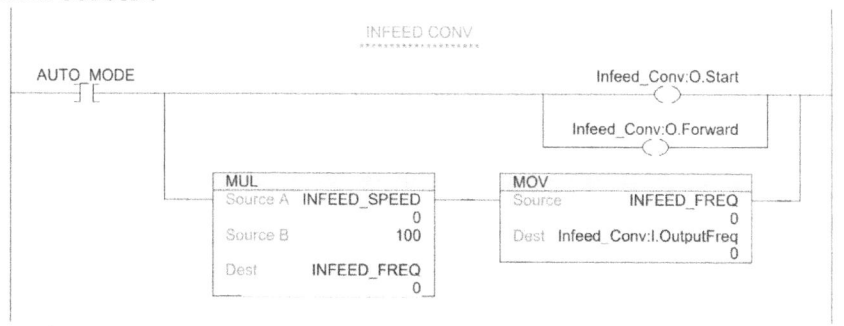

```
                          INFEED CONV

  AUTO_MODE                                                     Infeed_Conv:O.Start
   ─┤ ├─                                                              ─( )─
                                                              Infeed_Conv:O.Forward
                                                                     ─( )─
               ┌─ MUL ─────────────────┐   ┌─ MOV ─────────────────────┐
               │ Source A  INFEED_SPEED│   │ Source       INFEED_FREQ  │
               │                    0  │   │                       0   │
               │ Source B         100  │   │ Dest  Infeed_Conv:I.OutputFreq │
               │ Dest     INFEED_FREQ  │   │                       0   │
               │                    0  │   └───────────────────────────┘
               └───────────────────────┘
```

- When the roller conveyor is at bottom of the elevator and there is no any box present on it, the box will start loading on the conveyor.
- Once the conveyor reaches at the top of the elevator, the box will start unloading itself from the roller conveyor, and once the box passes on the outfeed conveyor, the roller conveyor will stop.

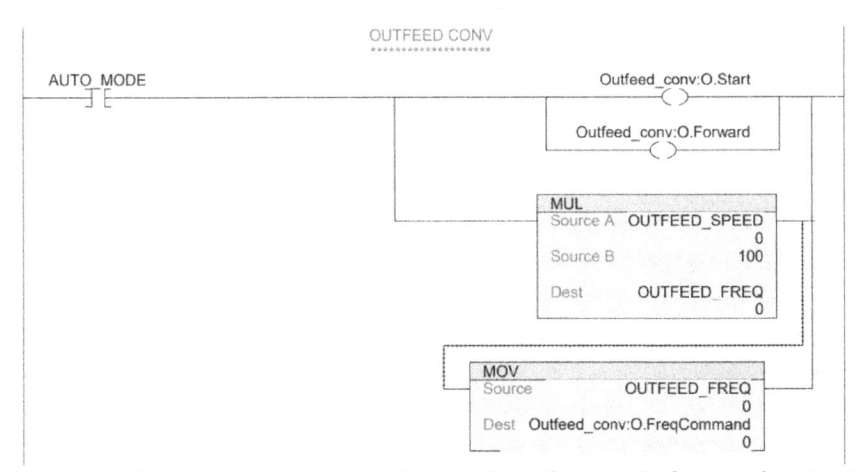

```
                                    OUTFEED CONV
AUTO_MODE                                               Outfeed_conv:O.Start
───┤ ├──────────────────────────────────────────────────────( )──────
                                                        Outfeed_conv:O.Forward
                                                              ( )
                                      ┌─────────────────────────────┐
                                      │ MUL                         │
                                      │ Source A  OUTFEED_SPEED     │
                                      │                        0    │
                                      │ Source B              100   │
                                      │ Dest      OUTFEED_FREQ      │
                                      │                        0    │
                                      └─────────────────────────────┘
                                      ┌─────────────────────────────┐
                                      │ MOV                         │
                                      │ Source    OUTFEED_FREQ      │
                                      │                        0    │
                                      │ Dest  Outfeed_conv:O.FreqCommand│
                                      │                        0    │
                                      └─────────────────────────────┘
```

- The roller conveyor speed can be changed from the HMI screen.

```
                              ROLLER CONVEYOR
                                         PRODUCT PRESENT
                     BOTTOM STOP SENSOR       SENSOR
AUTO_MODE                    I0.7              I1.3             PRODUCT_LOAD
───┤ ├──────────────────────┤ ├────────────────┤/├───────────────────(L)────
                            PRODUCT PRESENT
                               SENSOR
                                I1.3                             PRODUCT_LOAD
                               ─┤ ├──────────────────────────────────(U)────

                                  TOP STOP SENSOR
AUTO_MODE                              I1.1                    PRODUCT_UNLOAD
───┤ ├────────────────────────────────┤ ├────────────────────────────(L)────
                                  OUTFEED SENSOR
                                       I1.5                    PRODUCT_UNLOAD
                                      ─┤ ├───────────────────────────(U)────
```

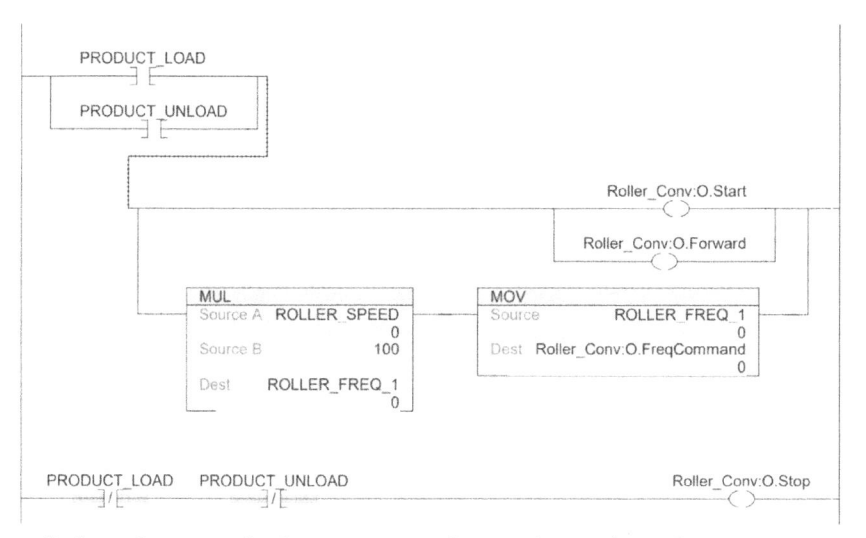

```
         PRODUCT_LOAD
         ─┤ ├─────────────────┐
         PRODUCT_UNLOAD        │
         ─┤ ├─────────────────┤
                               │                                    Roller_Conv:O.Start
                               └──────────────────────────────────────────( )──────────
                                                                    Roller_Conv:O.Forward
                                                  ┌───────────────────────( )──────────
                    ┌── MUL ──────────────────┐   │   ┌── MOV ─────────────────────┐
                    │ Source A  ROLLER_SPEED  │   │   │ Source          ROLLER_FREQ_1│
                    │                      0  ├───┤   │                           0  │
                    │ Source B          100   │   │   │ Dest  Roller_Conv:O.FreqCommand│
                    │ Dest     ROLLER_FREQ_1  │   │   │                           0  │
                    │                      0  │   │   └────────────────────────────┘
                    └─────────────────────────┘

   PRODUCT_LOAD    PRODUCT_UNLOAD                                       Roller_Conv:O.Stop
   ───┤/├──────────────┤/├─────────────────────────────────────────────────( )──────────
```

- If the elevator is in auto mode, and product is present on the roller conveyor, the elevator will start moving in upward direction.

- The elevator will start moving in high speed until it reaches top slow speed sensor and it will completely stop as it reaches top stop conveyor and the same thing happens when the elevator moves in downward direction.

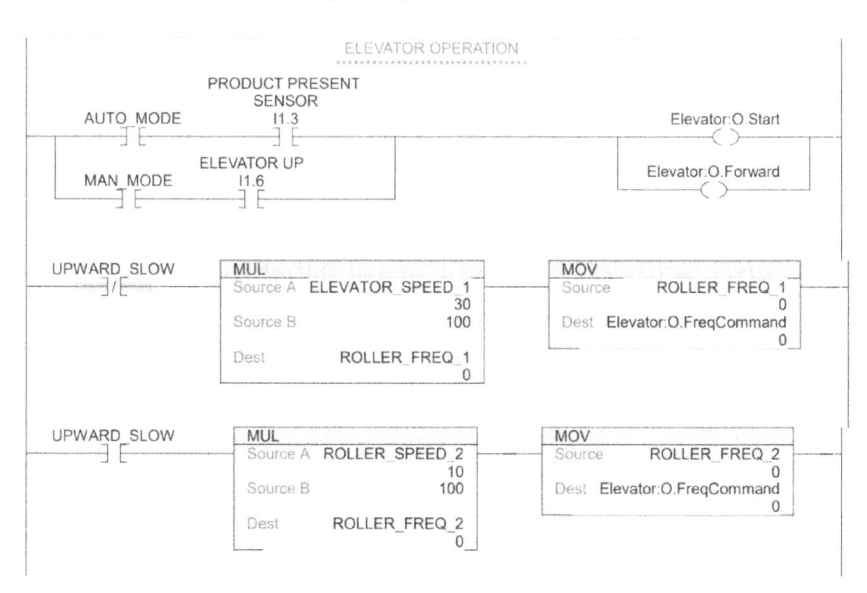

```
                        ELEVATOR OPERATION
                     PRODUCT PRESENT
                         SENSOR
      AUTO_MODE            I1.3                                    Elevator:O.Start
      ──┤ ├──────────────┤ ├──────────────────────────────────────────( )──────────
                     ELEVATOR UP                                   Elevator:O.Forward
      MAN_MODE           I1.6                                      ┌────( )────────┐
      ──┤ ├──────────────┤ ├─────────────────────────────────────────────────────────

   UPWARD_SLOW    ┌── MUL ──────────────────┐       ┌── MOV ─────────────────────┐
   ───┤/├─────────┤ Source A ELEVATOR_SPEED_1├──────┤ Source          ROLLER_FREQ_1│
                  │                     30  │       │                           0  │
                  │ Source B          100   │       │ Dest  Elevator:O.FreqCommand │
                  │ Dest     ROLLER_FREQ_1  │       │                           0  │
                  │                      0  │       └────────────────────────────┘
                  └─────────────────────────┘

   UPWARD_SLOW    ┌── MUL ──────────────────┐       ┌── MOV ─────────────────────┐
   ───┤ ├─────────┤ Source A  ROLLER_SPEED_2├───────┤ Source          ROLLER_FREQ_2│
                  │                     10  │       │                           0  │
                  │ Source B          100   │       │ Dest  Elevator:O.FreqCommand │
                  │ Dest     ROLLER_FREQ_2  │       │                           0  │
                  │                      0  │       └────────────────────────────┘
                  └─────────────────────────┘
```

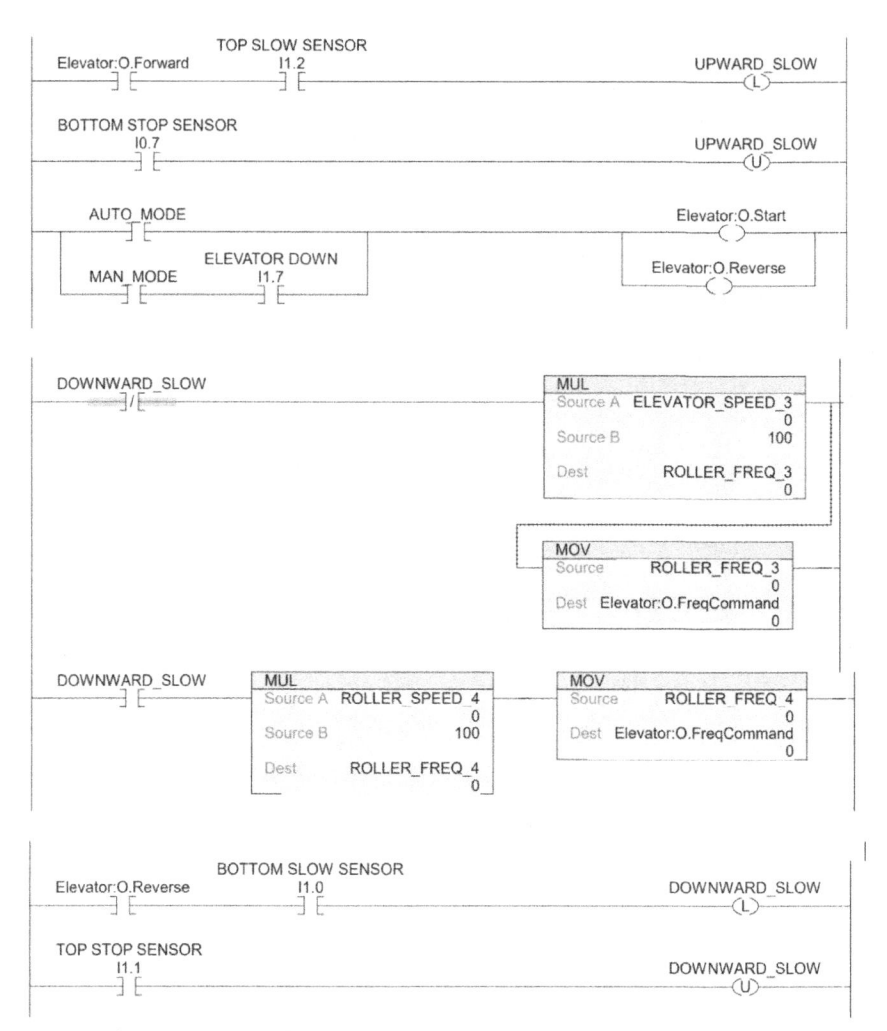

```
                    TOP SLOW SENSOR
  Elevator:O.Forward      I1.2                                    UPWARD_SLOW
  ──] [──────────────────] [──────────────────────────────────────────(L)──

  BOTTOM STOP SENSOR
        I0.7                                                      UPWARD_SLOW
  ──] [────────────────────────────────────────────────────────────────(U)──

       AUTO_MODE                                                 Elevator:O.Start
  ──] [──┬──────────────────────────────────────────────────────────────( )──
         │    ELEVATOR DOWN
       MAN_MODE    I1.7                                         Elevator:O.Reverse
  ──] [──┴──] [──────────────────────────────────────────────────────────( )──


   DOWNWARD_SLOW                        ┌─MUL─────────────────────┐
  ──]/[──────────────────────────────────┤ Source A  ELEVATOR_SPEED_3 ├──
                                         │                        0 │
                                         │ Source B             100 │
                                         │ Dest       ROLLER_FREQ_3 │
                                         │                        0 │
                                         └──────────────────────────┘
                                              ┌─MOV─────────────────────┐
                                              │ Source    ROLLER_FREQ_3 │──
                                              │                       0 │
                                              │ Dest Elevator:O.FreqCommand │
                                              │                       0 │
                                              └─────────────────────────┘

   DOWNWARD_SLOW      ┌─MUL─────────────────────┐   ┌─MOV─────────────────────┐
  ──] [───────────────┤ Source A  ROLLER_SPEED_4├───┤ Source    ROLLER_FREQ_4 │──
                      │                       0 │   │                       0 │
                      │ Source B            100 │   │ Dest Elevator:O.FreqCommand│
                      │ Dest   ROLLER_FREQ_4    │   │                       0 │
                      │                       0 │   └─────────────────────────┘
                      └─────────────────────────┘

                    BOTTOM SLOW SENSOR
  Elevator:O.Reverse      I1.0                                   DOWNWARD_SLOW
  ──] [──────────────────] [──────────────────────────────────────────(L)──

  TOP STOP SENSOR
        I1.1                                                     DOWNWARD_SLOW
  ──] [────────────────────────────────────────────────────────────────(U)──
```

Tower lamp operation

I have used a tower lamp on control panel that gives me the status of the elevator.

LEARN EVERYTHING ABOUT PLC PROGRAMMING

```
                          TOWER LAMP
                                                    TL GREEN
     AUTO_MODE                                        O0.0
  ─────] [─────────────────────────────────────────────( )─────

                                                    TL ORANGE
     MAN_MODE                                         O0.1
  ─────] [─────────────────────────────────────────────( )─────

                                                     TL RED
     SAFETY_OK                                        O0.2
  ─────]/[─────────────────────────────────────────────( )─────
```

3. SIEMENS PLC

A. Siemens PLC series

1. Logo
Perform small-scale automation tasks more quickly and free up space in your control cabinet: The LOGO! controllers from Siemens offer you the support you need at a price you can afford. With 8 basic logic functions and 30/35 special functions, LOGO! can replace a large number of conventional switching and control devices.

2. Logo 8
With LOGO 8, the successful logic module from Siemens starts the next generation. The new module meets almost all customer requests with easier handling, a new display and full communication options over Ethernet. It also makes the Web server application extremely easy to use.

Plus remote communication through wireless networks or a communication module rounds off the range of new opportunities associated with LOGO.

3. S7-200
For small applications s7-200 can be used. But this Controller are no more available for sale

4. S7-1200
The SIMATIC S7-1200 is the controller for control tasks in machine building and plant construction. The small controllers offer perfect interaction with the basic panels and are programmed with STEP 7 Basic in the TIA Portal.

5. S7-300
The S7-300 is the individual solution for fast process and automation tasks that contain additional data processing tasks. It is high-performance, fast, versatile, and future-proof. For engineering either STEP 7 V5.5/STEP, 7 Professional 2010 or STEP 7 Professional in the TIA Portal can be used.

6. S7-400

The SIMATIC S7-400, which is fast, robust, and strong in communication, has been proven as a controller in the upper-performance range of factory automation and can be found in plants for process automation as well.

Similar to the S7-300, for engineering, the established STEP 7, STEP 7 Professional in the TIA Portal or in the mighty, process-oriented PCS7 can be used.

7. S7-1500

The new SIMATIC S7-1500 controller sets new standards in productivity with its many innovations. SIMATIC S7-1500 is perfectly integrated with the TIA Portal for maximum efficiency in engineering.

The ultimate plus in automation is furthermore convincing through its system performance, its integrated technology, its security concept, easiest handling, and maximum usability and last but not least through its integrated system diagnostics with consistent visualization concept in the CPU display and in the engineering, on the HMI-Panel and in the web server.

8. S7-1500 Software Controller

The S7-1500 Software Controller CPU 1507S implements the function of an S7-1500 controller as software on a SIMATIC IPC with Windows. This allows a SIMATIC IPC to be used for control of machinery or plants. The integrated Ethernet and PROFIBUS interfaces of the SIMATIC IPC can be used to connect distributed I/O via PROFINET or PROFIBUS.

In addition, the CPU offers extensive control functions through easily configurable modules, as well as the connection of drives via standardized PLC-open blocks. Its special strengths as a software controller come into play when specific automation functions are integrated via the programming language C or C++, or when a close connection between Windows software and the software controller is required.

9. Simatic ET 200SP

The SIMATIC ET 200SP distributed I/O system is a scalable

and highly flexible, distributed I/O system in the degree of protection IP20 for linking of process signals to a central controller via PROFINET.

10. Simatic ET200S

The ET 200S is a modular, distributed I/O device in the degree of protection IP 20 with integrated SIGUARD safety technology. In addition to input/output modules, the ET 200S integrates technology modules and load feeders.

11. Simatic ET200 Pro

SIMATIC ET 200pro is a modular, rugged and versatile I/O system with IP65/66/67 degree of protection. It comprises Interface Modules to connect with PROFINET or PROFIBUS environments with standard as well as failsafe functionality. Interface modules are also available with CPU functionality.

Due to its rugged design ET 200pro can be used when mechanical load is high. Hot swapping of electronics and terminal modules during operation increases plant availability.

B. Communication protocols

1. Profinet

This communication protocol is mainly used by siemens PLC.Profinet is an industrial ethernet-based system.Using Profinet we can connect facilities like PLC, HMI, Distributed I/O, a different type of transmitters, sensor, actuator, VFD, etc. all on one network. Profinet provides faster response time so the collection of the data becomes even greater. Industrial Profinet comes with shielding provides better performance in an electrically noisy environment.

To recognize Profinet devices it must be defined with IP address and device name.Each device in a network must be assigned with IP address to communicate. Profinet cables are easily recognizable by their green color.Profinet operates at 100 megabits per second, and cables may be up to 100 meters in length.Due to its high-speed operation and a response time of less than 1 millisecond, Profinet is ideal for high-speed applications.Because Profinet uses the same physical connection standards as Ethernet, standard Ethernet switches can be used to expand your network.

2. Profibus

A Profibus connection port may look very familiar to you; It looks just like a standard DB-9 serial connector. While it may look the same, the underlying protocol is very different. It is easily recognizable by its purple outer jacket.Profibus networks operate at speeds of 9600 bits per second to 12 megabits per second. While Profibus cables may be up to 1000 meters long, shorter cable lengths are required for higher data rates.

C. Siemens PLC programming software

SIMATIC STEP 7

Programming languages: LAD, FBD, SCL, STL, S7 Graph

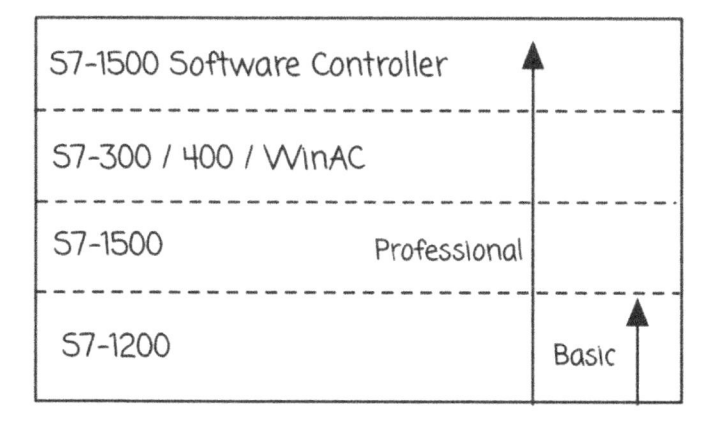

S7-1500 Software Controller	
S7-300 / 400 / WinAC	
S7-1500	Professional
S7-1200	Basic

```
┌─────────────────────────────────┐
│         SIMATIC WinCC           │
└─────────────────────────────────┘

┌─────────────────────────────────┐
│  Machine level operation monitoring │
│                                 │
│  SCADA applications             │
└─────────────────────────────────┘

┌─────────────────────────────────┐
│  SCADA              Professional│
│  ---------------------------    │
│  Single user PC    Advanced     │
│  ---------------------------    │
│  Comfort panel   Comfort        │
│  ---------------------------    │
│  Basic panel                    │
│                  Basic          │
└─────────────────────────────────┘
```

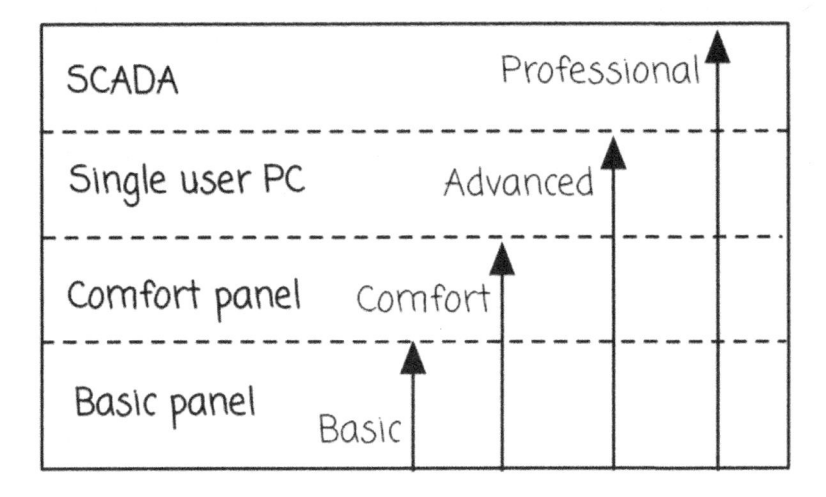

1. TIA Portal

The Totally Integrated Automation (TIA) Portal is Siemens' flagship programming software for its PLCs. It supports all of the Siemens PLC series and provides a centralized platform for programming, simulation, and diagnostics.

You do not need extra software like Allen Bradley for HMI programming and communication.

Programming languages:

- Ladder logic
- Structured Text
- Functional block

- Sequencial flow chart

2. SIMATIC STEP 7

This is another programming software for Siemens PLCs. It supports the SIMATIC S7-1200, S7-1500, and S7-300 PLC series, and offers a wide range of programming languages, including ladder logic, function block diagram, and structured text.

3. SIMATIC WinCC

This is a visualization software that can be used with Siemens PLCs. It provides an intuitive interface for monitoring and controlling the PLC, and supports a wide range of communication protocols.

4. SIMATIC ProTool

This is a legacy visualization software that can be used with older Siemens PLCs. It provides a simple interface for monitoring and controlling the PLC, and supports multiple languages.

3.1 TIA PORTAL

A. Introduction

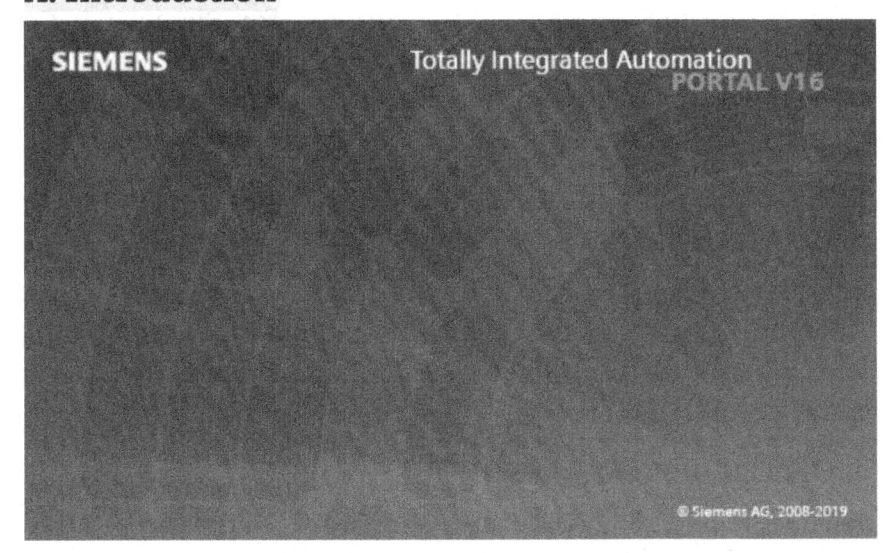

TIA Portal is a software and tools package developed by Siemens, which aims to integrate multiple development tools for automation devices from the unification and remodelling of preexisting software such as Simatic Step 7, Simatic WinCC, and Sinamics Starter.

The environments are responsible for programming, developing, and configuring Siemens PLCs, HMIs, and frequency inverters. The user's programming logic in TIA Portal follows a structure of blocks, a facilitating agent for the development, maintenance, and diagnostics of machines and industrial processes when developed in a structured and organized way.

Step7 is used to program PLCs of the S7-1200, S7-1500, S7-300, and S7-400 families. WinAC and the latest S7-1500 Software Controller are alternative controllers for industrial computers.

The available programming languages are ladder, FBD (Function Block Diagram), SCL (Structured Control Language),

STL (Statements List), and S7 GRAPH. The Development of HMI screens in WinCC is applied to supervisory systems on computers, isolated or SCADA, and Basic, Comfort, and Mobile operational panels. Profibus, PROFINET, and AS-I (Actuator Sensor Interface) communication protocols.

For communication with PLCs, it is worth mentioning the existence of CMs (Communication Modules) with functions to establish communications in different industrial protocols, such as Modbus and CANOpen. The Figure below gathers the main features for SIMATIC STEP 7 and SIMATIC WinCC.

1. Portal view

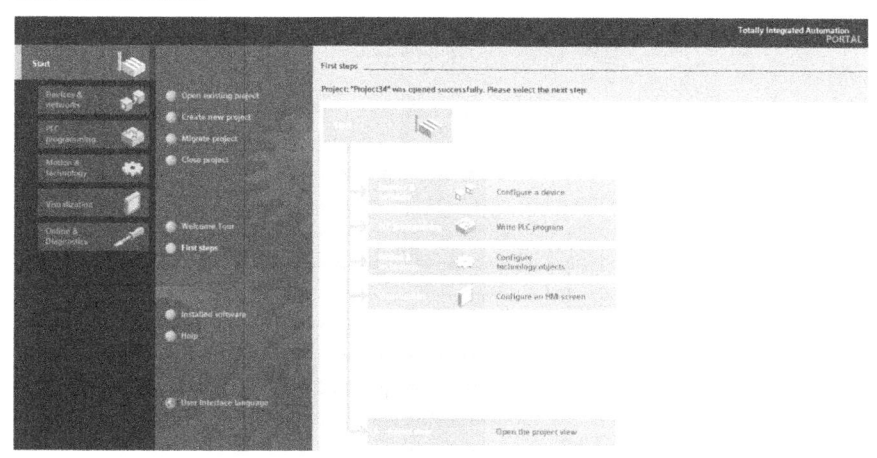

2. Project view

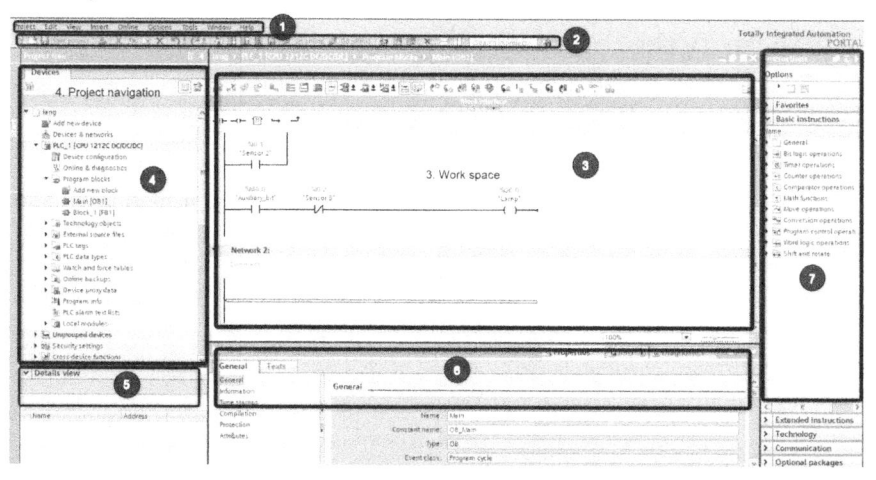

1. Menu bar: The menu bar contains all the commands that you require for your work.

2. Toolbar: The toolbar provides you with buttons for commands you will use frequently. This gives you faster access to these commands than via the menus.

a. Upload and download a program

1. Program download- The marked icon is used to download a program from PC to PLC.

2. Program upload- The marked icon is used to upload a program from PLC to PC.

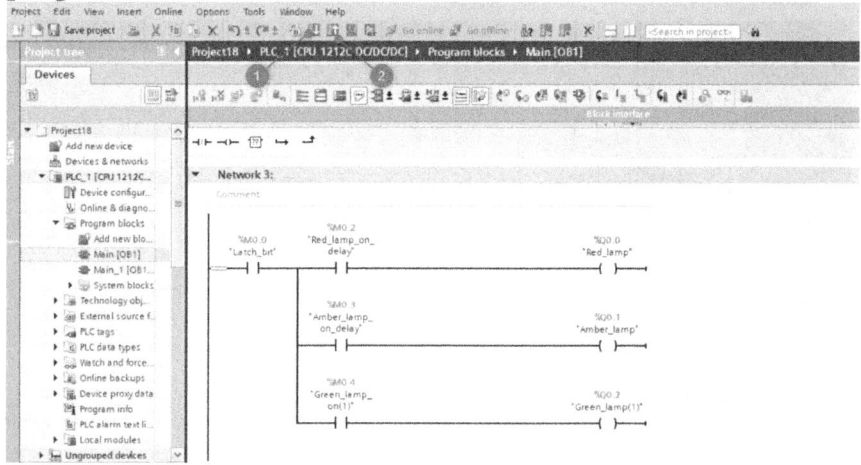

While doing any one options a screen will appear in front of you to verify the communication.

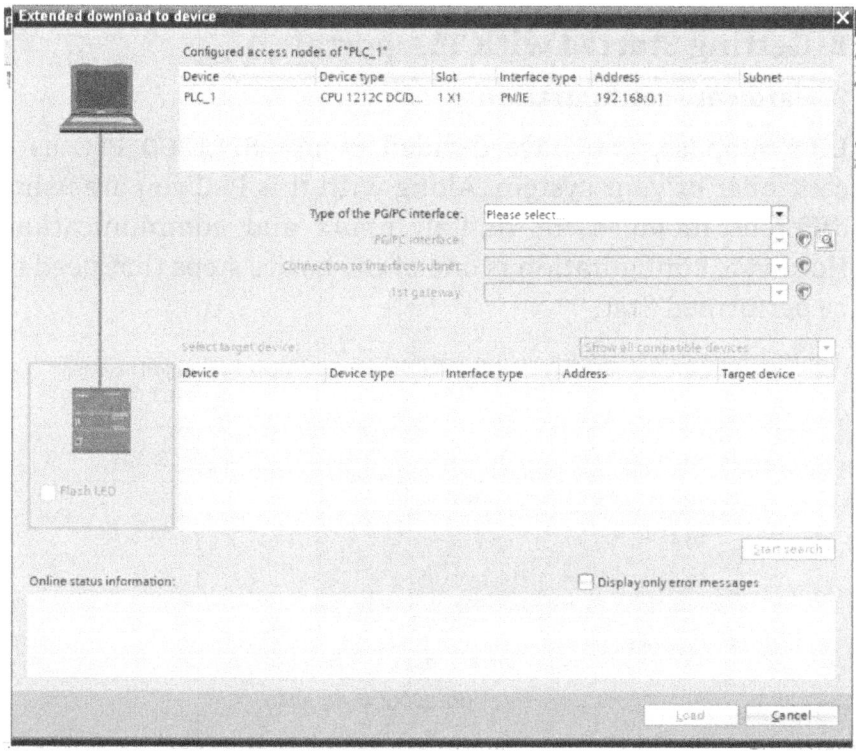

3. Project tree: The project tree gives you access to all components and project data. You can perform, for example, the following tasks in the project tree:

• Add new components
• Edit existing components
• Scan and modify the properties of existing components

4. Work area: The objects that you can open for editing purposes are displayed in the work area.

5. Task cards: Task cards are available depending on the edited or selected object. The task cards available can be found in a bar on the right-hand side of the screen. You can collapse and reopen them at any time.

6. Details view: Certain contents of a selected object are shown in the details view. This might include text lists or tags.

B. Getting Started with TIA portal

2. Hardware configuration

Let's suppose, you have decided to use S7 1200 PLC as a controller of your system. Along with this PLC you are using different modules for digital inputs and communication. Hardware configuration is one of the basics steps that need to be performed first.

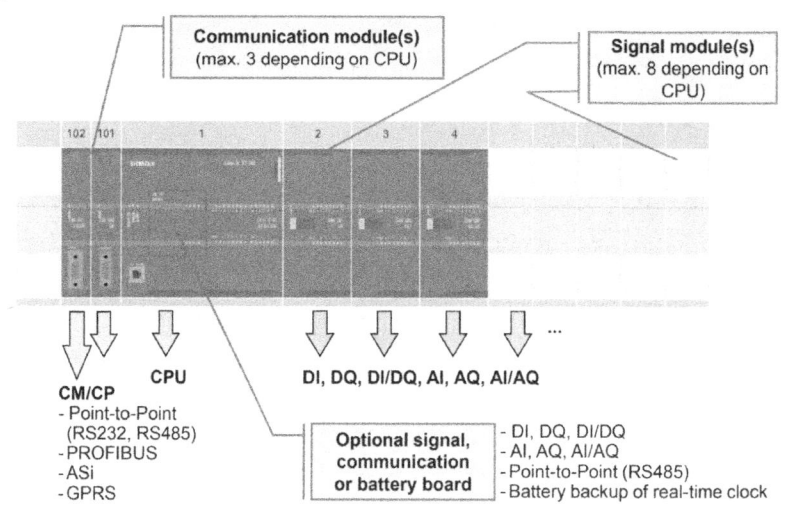

In this section I am going to show you the configuration of these hardware during PLC programming.

Select a controller

When you open TIA portal, this screen will appear in front of you. You select controller, HMI, etc. from there.

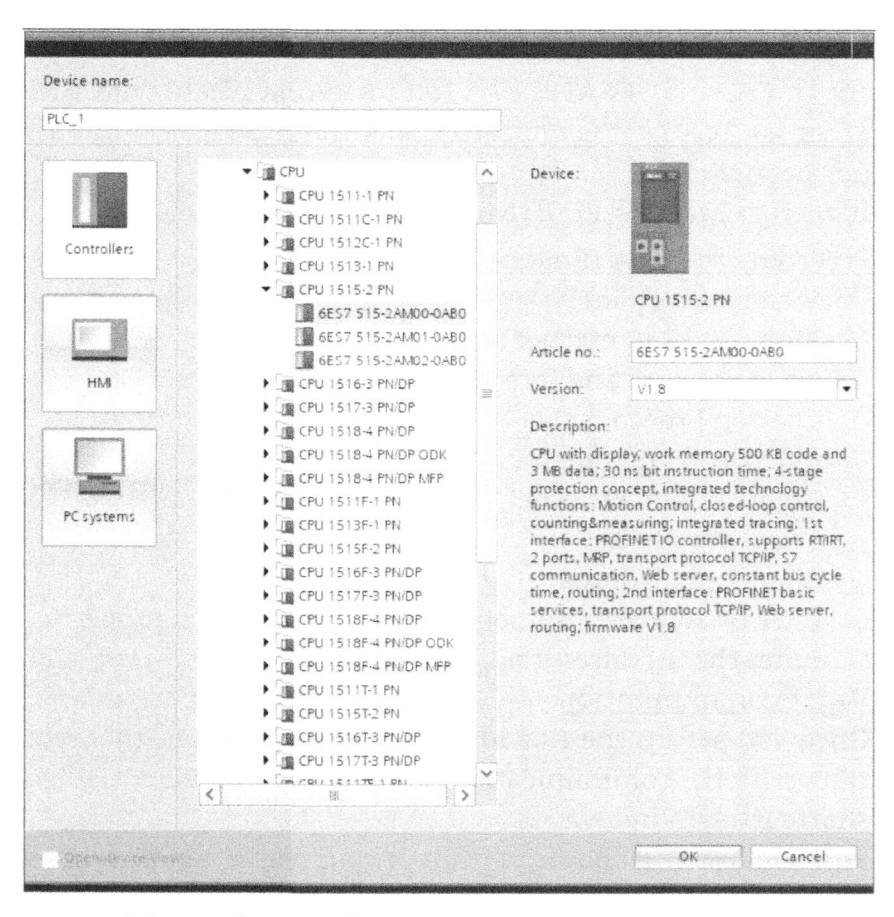

Set IP address of controller

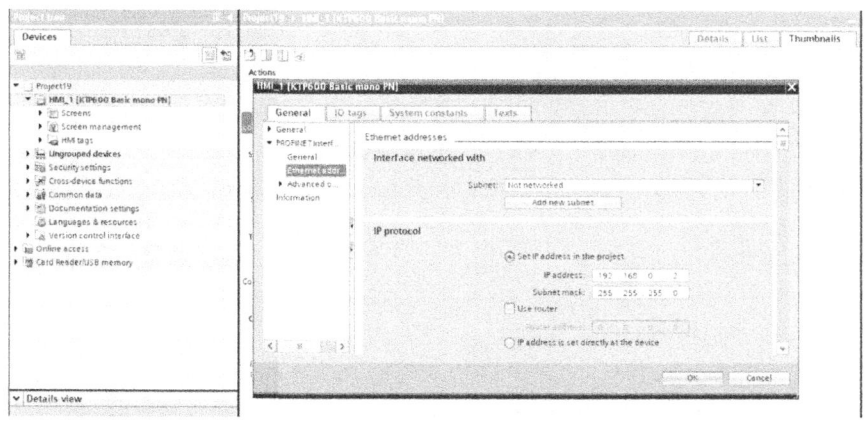

To program the SIMATIC S7-1200 controller from the PC, the programming device or a laptop, you need a TCP/IP connection or an optional PROFIBUS connection.

For the PC and SIMATIC S7-1200 to communicate with each other via TCP/IP, it is important that the IP addresses of both devices match.

Property>General> PROFINET interface> IP Protocol

Procedure to set up IP address of PC
- Visit control panel
- Right click on internet access
- Open network and internet settings
- Click on Ethernet
- Click on change adapter setting
- Right click on Ethernet
- Select properties
- Simply scroll down
- Click on Internet protocol Version 4, then select property
- Enter the IP address and, then click on OK.

Verify communication

Once you set up the IP address of your computer, then you can verify the communication between PLC and PC by using a shortcut key Windows+R.

Run window will appear in front of your computer, then enter the IP address you want to ping. For e.g. ping 192.168.10.1-t

If your connection is OK, then your PC will get response from PLC.

Add a module

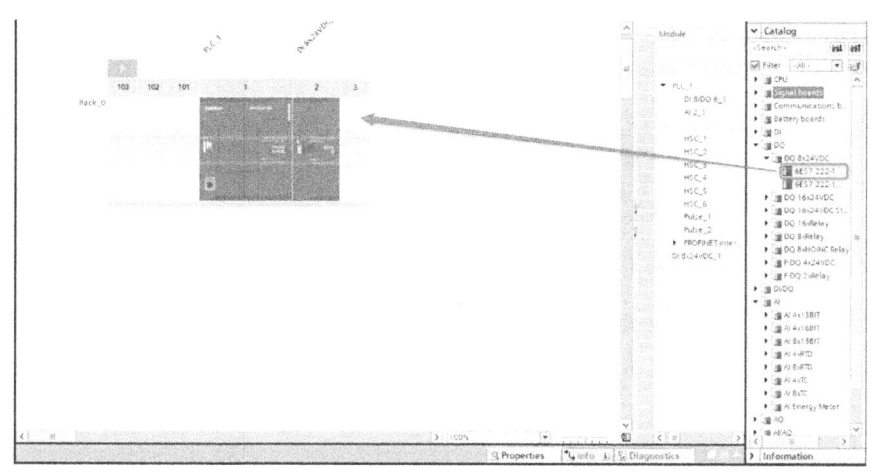

Sometimes as per the application you need to add different modules with the controller. In that case you cannot just attach them physically with PLC and leave. Your work is incomplete until you configure it through software also Adding a module is not so difficult, just drag and drop.

The list of modules can be provided with PLC:

- DI Module
- DQ Module
- DI/DQ Module
- Communication module
- AI Module
- AQ Module
- AI/AQ Module
- Technology module

Change IO address

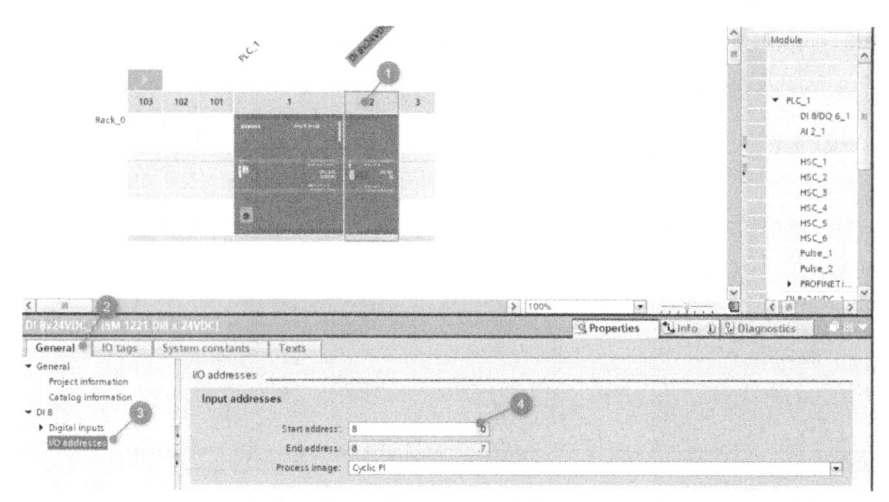

You are to parameterize the I/O addresses of the integrated Inputs and Outputs of the CPU as shown in the picture.
1. Select the module
2. Go to general
3. Select IO addresses
4. Enter the IO address and click OK

Configuration of peripheral devices

1. Configuration of module using GSD file

a. Download GSD file of a module

b. Option> Manage general station description

c. Give a source path

d. Install GSD file

e. Search with module number

Compiling the Device Configuration

1. Right click on PLC
2. Select compile
3. Hardware and software compile

LEARN EVERYTHING ABOUT PLC PROGRAMMING

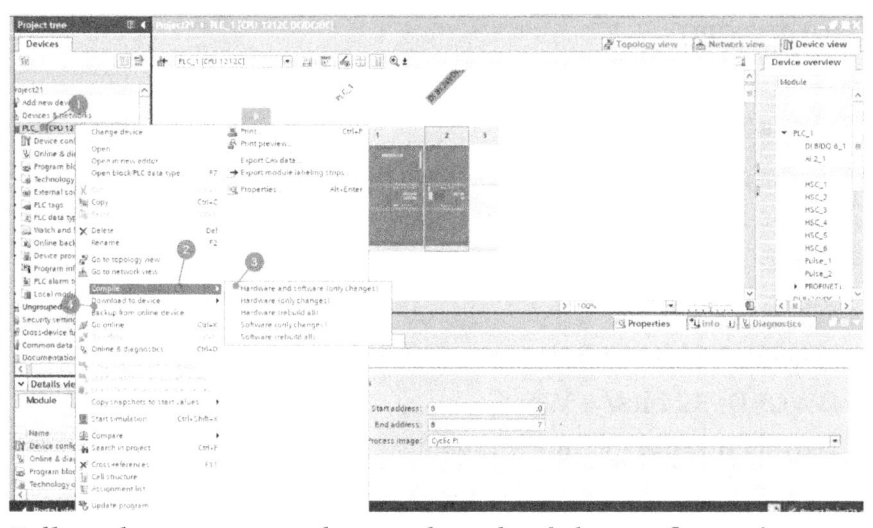

Follow the same procedure to download the configuration.

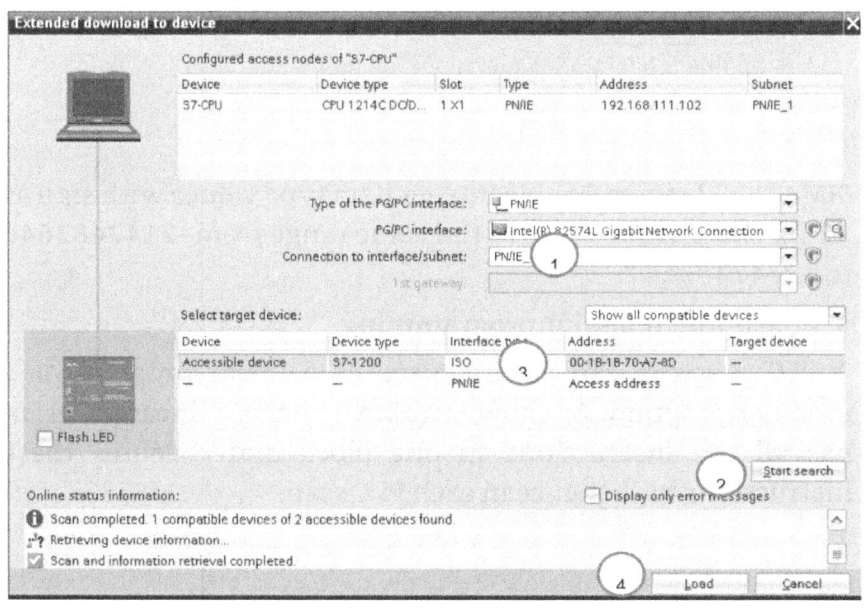

Fill in the dialog in the order which is shown in the picture above.

1. In "Connection to interface/subnet" select PN/IE
2. For refreshing the list "Show all compatible devices" activate "Start search"
3. In refreshed list "Show all compatible devices" select the

S7-1200 CPU

4. Start the download using the "Load" botton.

3. Data Type

The data type specifies how the value of a tag or constant is to be used in the user program.

- BOOL- 0/1
- BYTE- 8 bits
- WORD- 16 bits/ 2 Byte
- DWORD- 32 bits/ 2 Word
- Double Integer (32-Bit Integer)

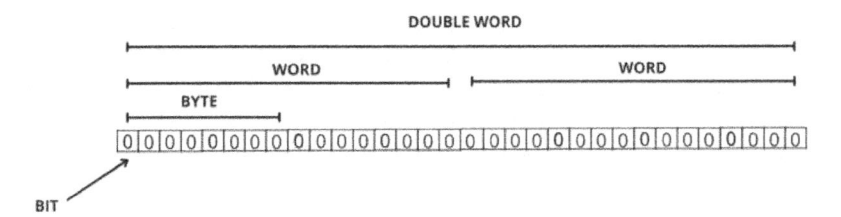

```
                    DOUBLE WORD
          WORD                    WORD
    BYTE
   |0|0|0|0|0|0|0|0|0|0|0|0|0|0|0|0|0|0|0|0|0|0|0|0|0|0|0|0|0|0|0|0|
BIT
```

SIMATIC S7 stores Double Integer data type values with sign as 32-bit code. This results in the value range from -2147483648 to +2147483647.

4. Linear and modular programming

A PLC program can be organized as a linear program or a modular program. As shown in above image, a linear program has all the instructions in one block and executes these instructions in sequence in each PLC scan.

a. Linear programming   b. Modular programming

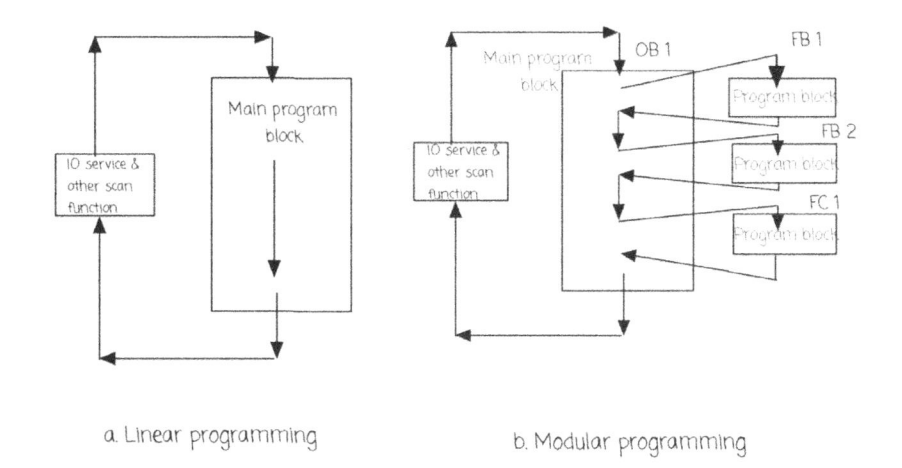

A modular program is composed of multiple program blocks. Some program blocks are executed in each PLC scan and other program blocks are executed under special circumstances. The program blocks can also be nested. This means that one program block can call another program block which can call another program block. This provides another flexibility in programming.

5. Types of program blocks

The controller programming is based on an architecture segmented into the blocks OBs (Organization Blocks), FCs (Functions), FBs (Function Blocks), and DBs (Data Blocks).

Blocks -The automation system provides various types of blocks in which the user program and the related data can be stored. Depending on the requirements of the process, the program can be structured in different blocks. You can use the entire operation set in all blocks (FB, FC and OB).

Organization Blocks (OBs) Organization blocks (OBs) form the interface between the operating system and the user program. The entire program can be stored in OB1 that is cyclically called by the operating system (linear program) or the program can be divided and stored in several blocks (structured program).

Functions (FCs) A function (FC) contains a partial functionality of the program. It is possible to program functions as parameter-assignable so that when the function is called it can be assigned parameters. As a result, functions are also suited for programming frequently recurring, complex partial functionalities such as calculations.

Function Blocks (FBs) Basically, function blocks offer the same possibilities as functions. In addition, function blocks have their own memory area in the form of instance data blocks. As a result, function blocks are suited for programming frequently recurring, complex functionalities such as closed-loop control tasks.

Data blocks are used for storing user data. They occupy memory space in the user memory of the CPU. Data blocks contain variable data (such as numeric values) with which the user program works. The user program can access the data in a data block. Access can be made symbolically or absolutely.

Working of Blocks

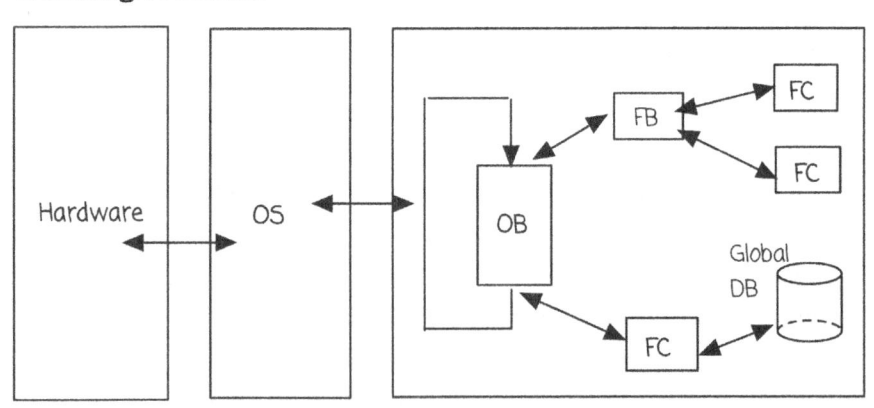

Organizational blocks are closely linked to program execution cycles and interrupts, as their execution is linked to previously defined triggers, such as a time interval or hardware failure detection. The OB1, or simply main, is essential for the initialization and sequential scanning of the calls of the blocks corresponding to the implemented code, except for other OBs, because it is a cyclic and continuous execution block. According to the representation of the Figure, the program execution starts with the contents of OB1 linearly and synchronously, codes from left to right and from the beginning to the end of the block. At the end of the main, the CPU resumes the execution of the code from the start and this process repeats indefinitely.

Although the user creates several blocks of functions and data, the PLC will only execute the dictated instructions and in the stipulated sequence through OB1. The Figure below demonstrates the creation of subprograms as evidenced by the block calls, which may have nested calls.

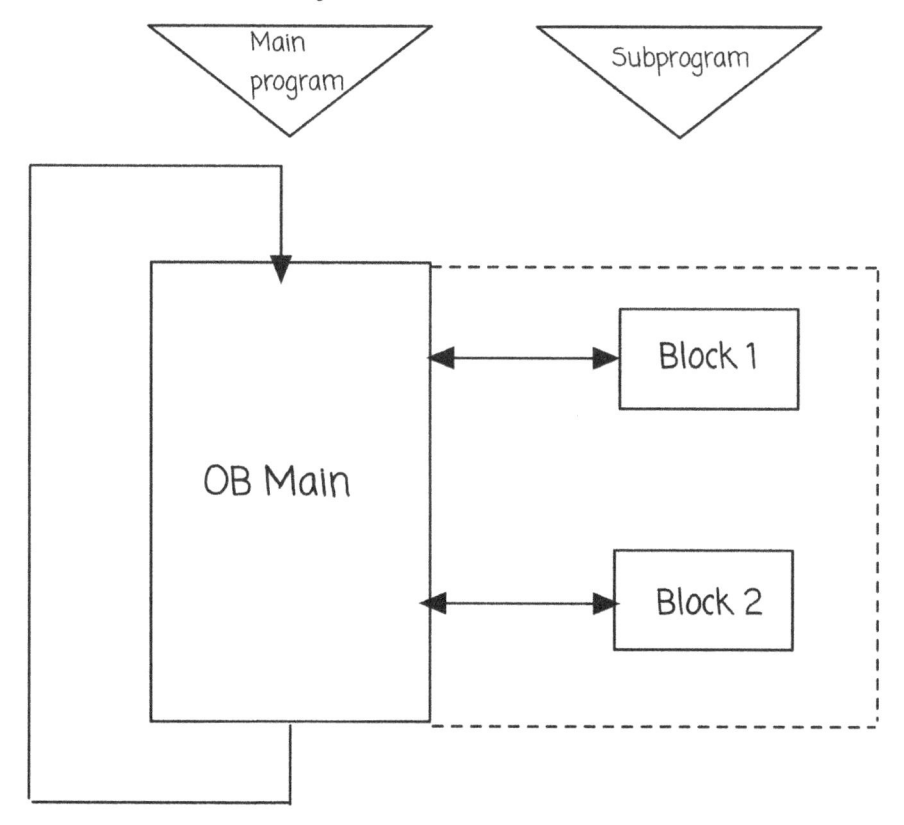

Since the reproduction of instructions in the main function depends on the processing of all assigned functions, its cycle time may vary with the state of the plant and any unanticipated programming conditions. Therefore, to avoid logical crashes or to guarantee the cyclical execution with a continuous activation period of a certain logical treatment, the Cyclic Interruption OB is recommended.

Cyclic Interrupt blocks are also executed cyclically but with a predefined time interval between executions when creating the block.

Thus, every 100ms, the current processing is interrupted so that the instructions contained in the Cyclic Interrupt block are carried out.

C. TIA Ladder logic Instructions

1. Bit logic operations

NO/NC instructions

A Digital signal is that signal which is in a discrete manner. A digital signal could be either 0Vdc or 24V dc supply.

Addressing

An address exactly defines where values are written or from where they are read. They begin at byte address, that is, at the number "0". The addresses are consecutively numbered, and, within a byte, the bit address 0..7 is numbered from right to left.

The % character identifies the presentation of an address.

PLC Tags

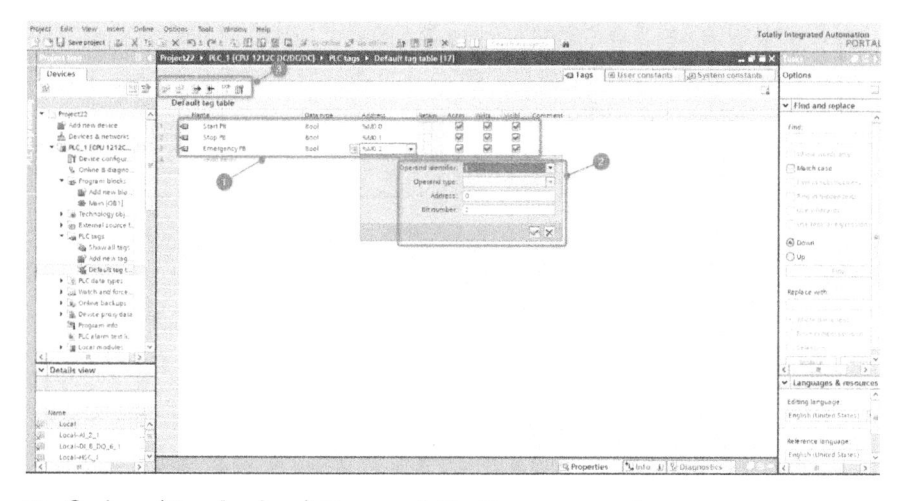

Defining (Declaring) Tags while Programming

If, during programming, unknown tags are used, they can be defined later-on network-by-network. Advantage: In the dialog that appears, only addresses are offered that are unused up to that point. In this way, for example, errors are avoided such as the using of bits that belong to an already used word (overlapping access).

While defining a tag, you have to select the types of tags whether it is global input, memory or output bool.

<u>After defining all the tags</u>

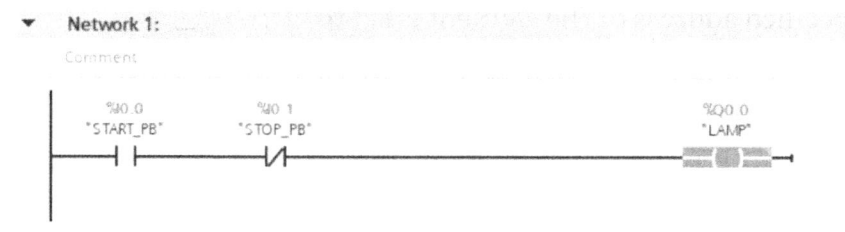

Rewire Tags while Programming

TIA provides with a facility to edit the existing PLC tags. Simply right click on the tags you want to edit and select rewire tag.

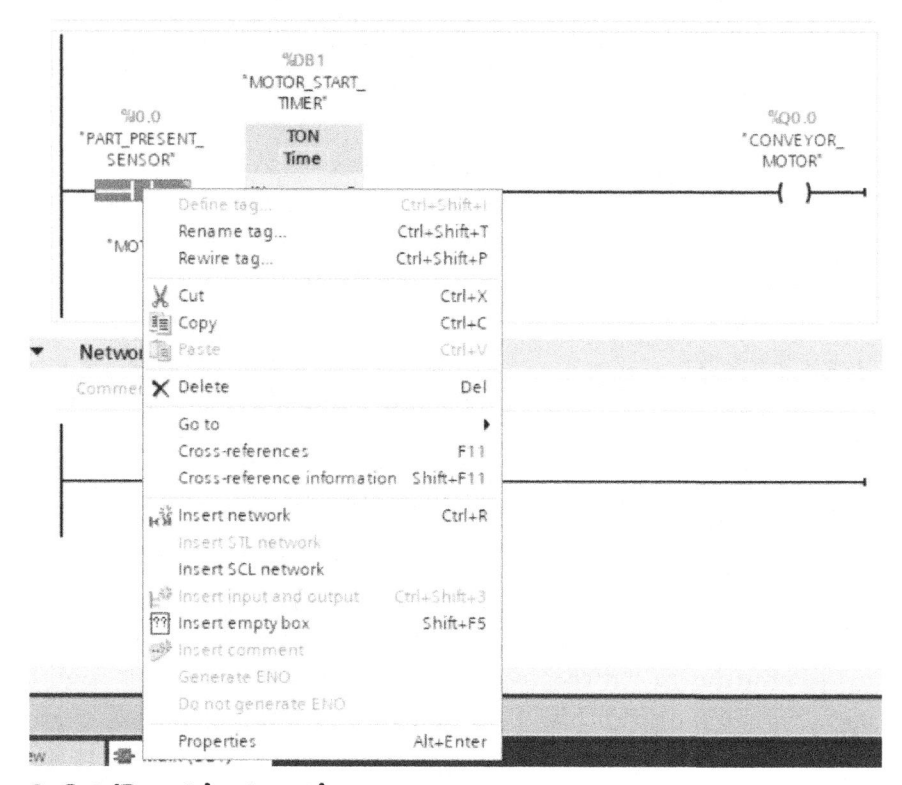

2. Set/Reset instructions

Set coil is executed only if the RLO of the preceding instructions is "1" (power flows to the coil). If the RLO is "1" the specified address of the element is set to "1".

An RLO=0 has no effect and the current state of the element's specified address remains unchanged.

Reset coil is executed only if the RLO of the preceding instructions is "1" (power flows to the coil). If power flows to the coil (RLO is "1"), the specified <address> of the element is reset to "0". A RLO of "0" (no power flow to the coil) has no effect and the state of element's specified address remains unchanged.

The address may also be a timer whose timer value is reset to "0" or a counter whose counter value is reset to "0".

Tower lamps control

- When switch 1 is pressed, lamps 1,2, and 3 get on,
- When switch 2 is pressed, lamps 4 and 5 get on,
- When switch 3 is pressed, lamps 1, and 5 will get off.

Input devices

- Switch 1 (I0.0)
- Switch 2 (I0.1)
- Switch 3 (I0.2)

Output devices

- Lamp 1 (Q0.0)
- Lamp 2 (Q0.1)
- Lamp 3 (Q0.2)
- Lamp 4 (Q0.3)
- Lamp 5 (Q0.4)

```
%I0.0                                              %M0.0
"Switch_1"                                         "Set_bit1"
——| |————————————————————————————————————————————————( S )——

%I0.1                                              %M0.1
"Switch_2"                                         "Set_bit2"
——| |————————————————————————————————————————————————( S )——

%I0.2                                              %M0.2
"Switch_3"                                         "Set_bit3"
——| |————————————————————————————————————————————————( S )——
```

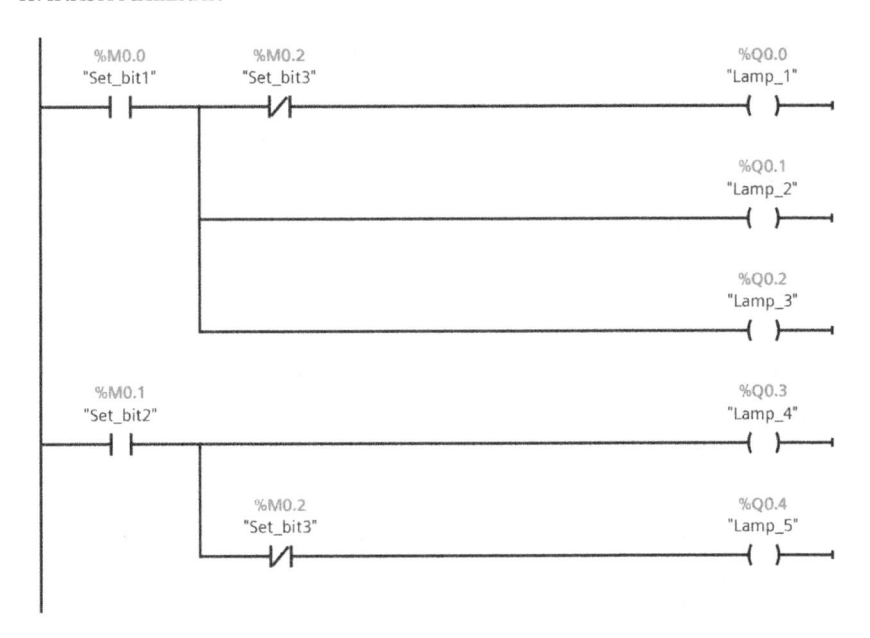

3. Positive and negative edge instructions

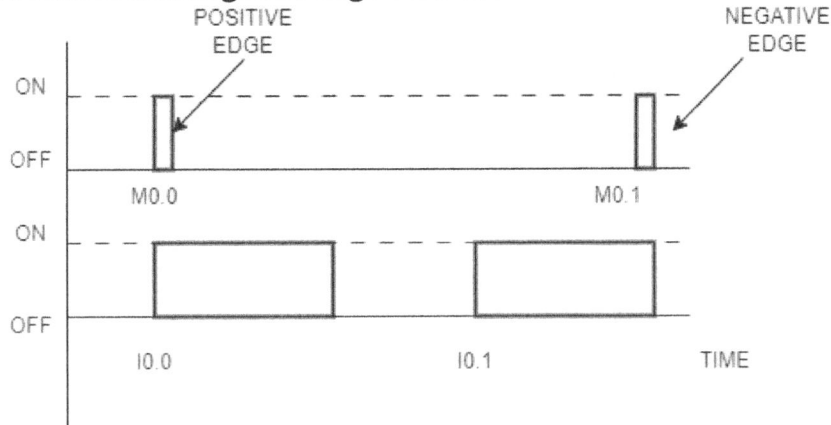

- The state of this contact is true when a positive transition (OFF to ON) is detected on assigned bit I0.0.

- The bit M0.0 will be ON for 1 clock cycle.

- The state of this contact is true when a negative transition (ON to OFF) is detected on assigned bit I0.0.

- The bit M0.0 will be ON for 1 clock cycle. These bits can be connected anywhere except output.

4. P Trig and N Trig

- The Q output power flow or logic state is true when a positive transition (OFF to ON) is detected on CLK power flow in ladder.
- The Q output power flow or logic state is true when a negative transition (ON to OFF) is detected on CLK power flow in ladder.

5. Timer operation

Basically there are four types of timers used in TIA portal.

a. TP (Generate Pulse)

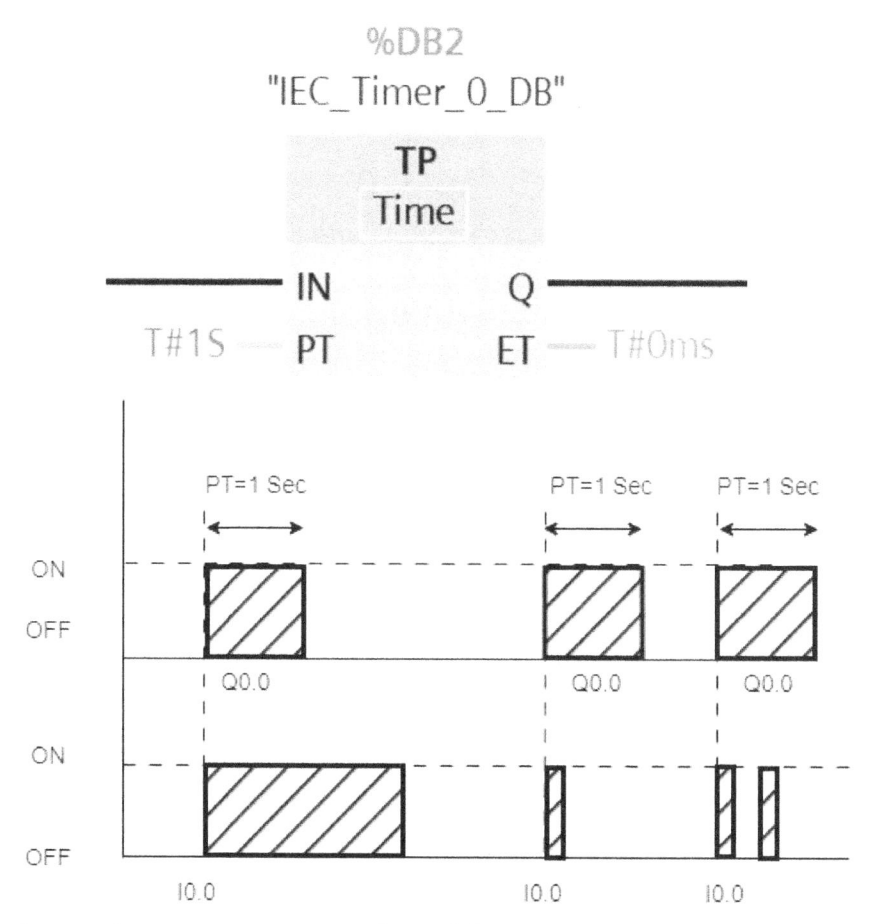

The TP timer generates a pulse with a preset width time.

b. TON (On Delay Timer)

```
            %DB2
         "IEC_Timer_0_DB"
            ┌─────────┐
            │   TON   │
            │  Time   │
       ─────┤IN      Q├─────
       T#1S │PT     ET│ T#0ms
            └─────────┘
```

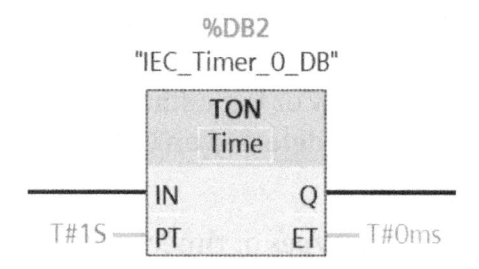

The TON timer sets the output (Q) to ON after a preset time delay.

c. TOF (Off Delay Timer)

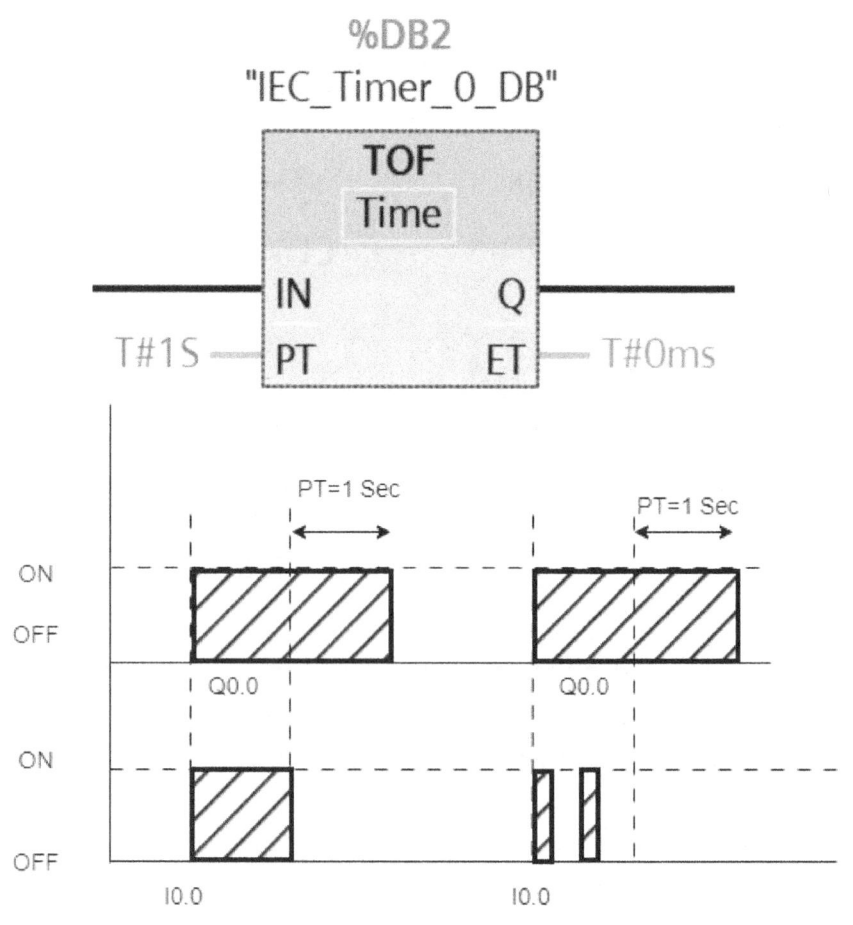

```
            %DB2
        "IEC_Timer_0_DB"
            ┌─────────┐
            │   TOF   │
            │  Time   │
       ─────┤IN      Q├─────
       T#1S │PT     ET│ T#0ms
            └─────────┘
```

- The TOF timer sets the output (Q) to ON and then resets the output to OFF after a preset time delay.

- If I0.0 sets from ON to OFF, the timer restarts again. So in order to operate this time, the I0.0 should be ON until the preset time gets over.

d. TONR (On Delay Retentive Timer)

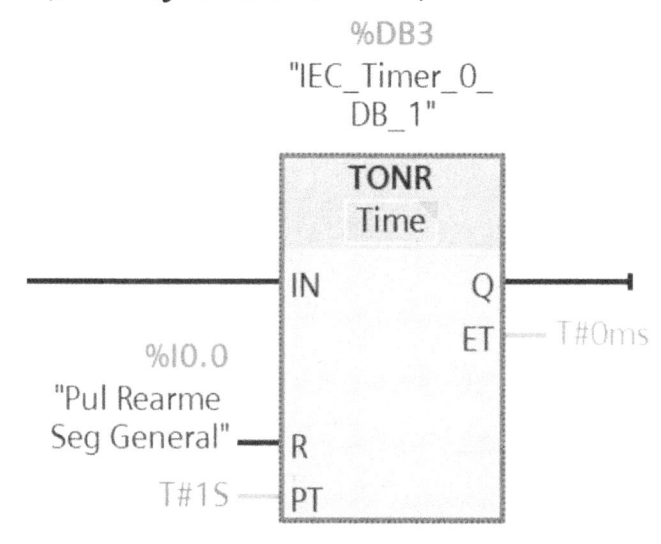

```
            %DB3
         "IEC_Timer_0_
            DB_1"
         ┌──────────┐
         │   TONR   │
         │   Time   │
─────────┤IN      Q ├──────
  %I0.0  │        ET├── T#0ms
"Pul Rearme │          │
Seg General"┤R         │
   T#1S ──┤PT        │
         └──────────┘
```

The TONR timer sets the output (Q) to ON after a preset time delay. The elapsed time is accumulated over multiple timing periods until the reset (R) input is used to reset the elapsed time.

e. PT (preset timer)

This coil loads a new preset time value in the specified timer.

f. RT (reset timer)

The RT coil resets the specified timer.

1. Traffic light control Application

The traffic light has red, amber, and green lamps. A red lamp gets on for 5 sec, then an amber lamp gets on for 2 sec and finally, the green lamp will get on for 5 sec. This cycle will continue for 1 minute.

Ladder Logic Diagram

Rung 1:
- `%I0.0` "Start_PB" —| |— ——(S)—— `%M0.0` "Latch_bit"

Rung 2:
- `%I0.1` "Stop_PB" —| |— ——(R)—— `%M0.0` "Latch_bit"

Rung 3:
- `%M0.0` "Latch_bit" —| |—
  - %DB1 "Red_lamp_timer" TON Time: IN, PT=T#5S, Q → `%M0.2` "Red_lamp_on_delay", ET=T#0ms
  - %DB2 "Red_lamp_off" TOF Time: IN, PT=T#5S, Q → `%M0.1` "Red_lamp_on", ET=T#0ms
  - `%M0.1` "Red_lamp_on" —| |— %DB5 "Amber_lamp_on" TON Time: IN, PT=T#2S, Q → `%M0.3` "Amber_lamp_on_delay", ET=T#0ms
  - %DB3 "Amber_lamp_off" TOF Time: IN, PT=T#7S, Q → `%M0.4` "Green_lamp_on(1)", ET=T#0ms
  - `%M0.4` "Green_lamp_on(1)" —| |— %DB4 "Green_lamp" TON Time: IN, PT=T#5S, Q → `%M0.5` "Green_lamp_on", ET=T#0ms

```
        %M0.0           %M0.2
      "Latch_bit"   "Red_lamp_on_                              %Q0.0
                         delay"                              "Red_lamp"
        ─┤ ├──────────────┤ ├─────────────────────────────────( )───

                         %M0.3
                     "Amber_lamp_                              %Q0.1
                       on_delay"                             "Amber_lamp"
                        ─┤ ├─────────────────────────────────( )───

                         %M0.4
                      "Green_lamp_                             %Q0.2
                         on(1)"                             "Green_lamp(1)"
                        ─┤ ├─────────────────────────────────( )───
```

2. Material conveying application

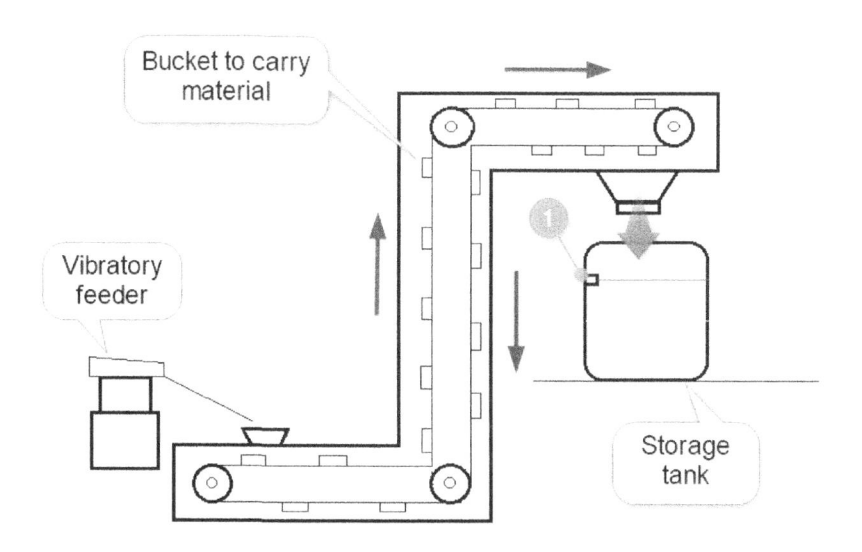

This type of material conveying system is used to carry material such as grains, powder, and small granules from ground level to a certain height. An electric motor is used to move the chain.

When a start PB is pressed, the chain starts rotating and the vibratory feeder starts after 2 sec. A vibratory feeder continuously feeds the material into the bucket. Then the material is unloaded into the storage tank. If the material reaches a level sensor, it will stop the vibratory feeder first and then the conveyor will stop after 2 sec.

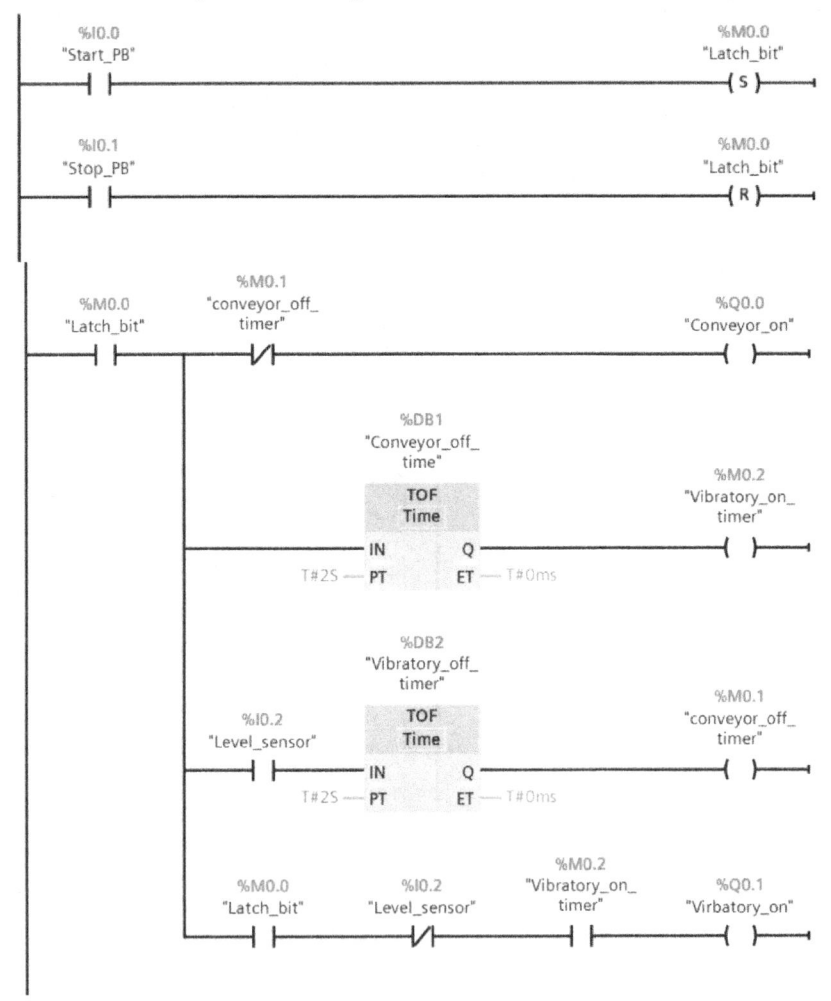

3. Box lifter

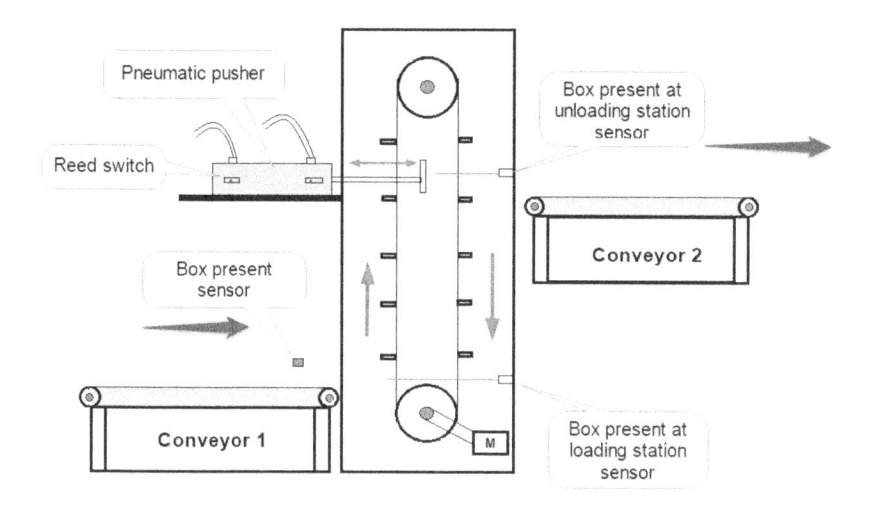

This material conveying system is used to lift the boxes and transfer them to a certain height.

When a start PB is pressed, the conveyor will start rotating. if a box is present at the conveyor 1 chain conveyor will stop. Then the box will enter into the conveyor, a box present sensor will sense the presence of the box at the loading station and the chain conveyor will start again after 1 sec.

As the box reaches the unloading sensor, the chain conveyor will stop. A pusher will actuate after 2 sec and it will push the box on the conveyor 2. After pushing the box, the pusher will retract and the reed switch will sense the pusher (homing sensor). Once the reed switch senses the pusher, the conveyor will start again after 1 sec. This cycle will continue until a stop PB is pressed.

4. Counter

Counters are used to count events, record quantities, etc. There are up counters and down counters as well as counters that can count in both directions.

Types of counter

There are three types of counter available in TIA portal.

a. CTU- Count up

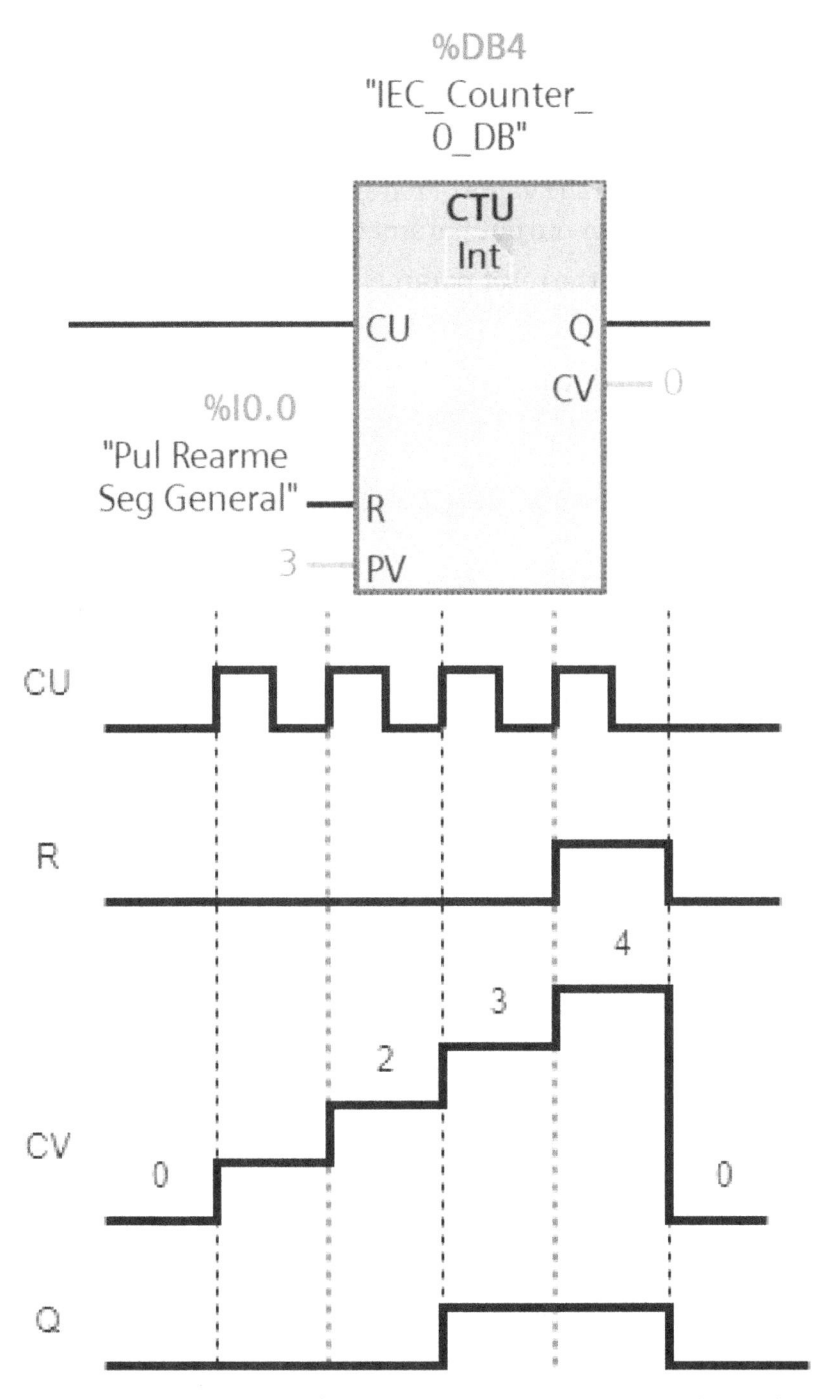

The "count up" counter (CTU) counts by 1 when the value

of the input parameter CU changes from 0 to 1. The timing diagram shows the operation of a CTU counter with an unsigned integer count value (where PV = 3).

- If the value of parameter CV (current count value) is greater than or equal to the value of parameter PV (preset count value), then the counter output parameter Q = 1.

- If the value of the reset parameter R changes from 0 to 1, then CV is reset to 0.

b. CTD- Count down

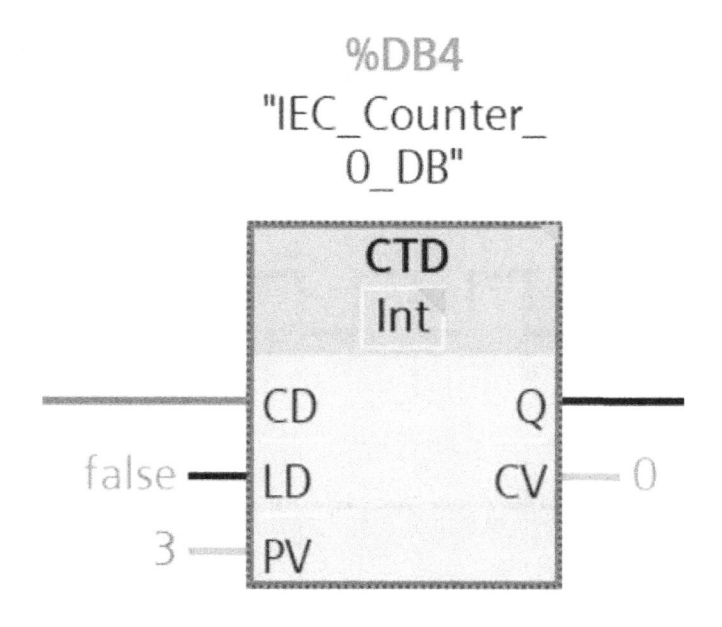

```
           %DB4
       "IEC_Counter_
           0_DB"

           ┌─────────────┐
           │    CTD      │
           │    Int      │
           │             │
    ───────┤ CD       Q  ├───────
  false ───┤ LD       CV ├─── 0
       3 ──┤ PV          │
           └─────────────┘
```

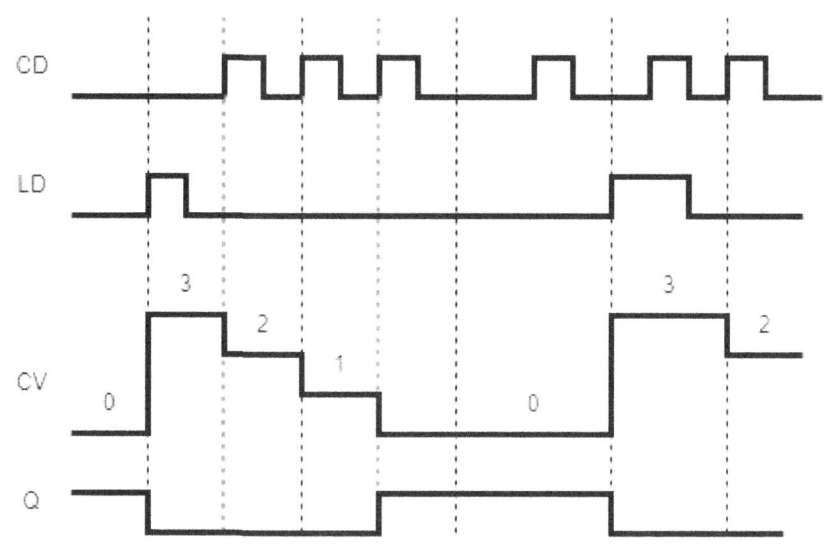

The "count down" counter (CTD) counts down by 1 when the value of input parameter CD changes from 0 to 1.

The timing diagram shows the operation of a CTD counter with an unsigned integer count value (where PV = 3).

- If the value of parameter CV (current count value) is equal to or less than 0, the counter output parameter Q = 1.

- If the value of parameter LD changes from 0 to 1, the value at parameter PV (preset value) is loaded to the counter as the new CV.

c. CTUD- Count up and down

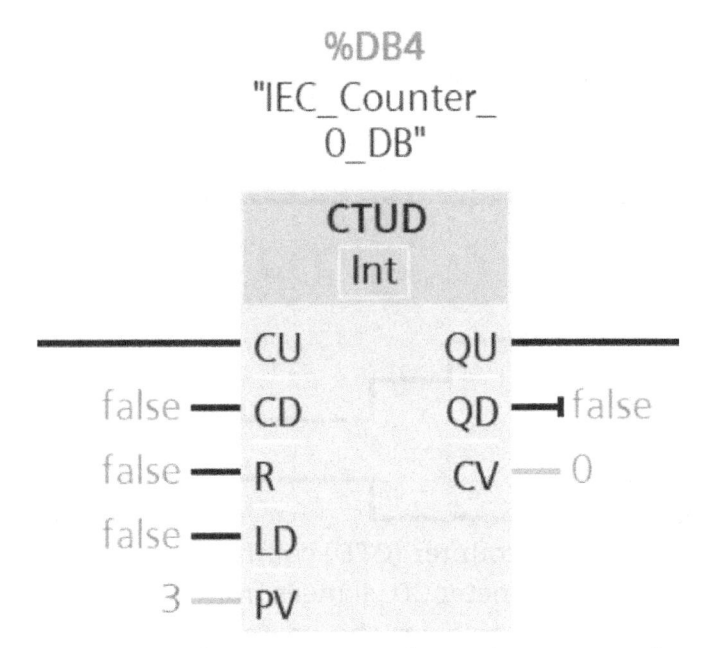

```
                %DB4
            "IEC_Counter_
                0_DB"
            ┌─────────────┐
            │    CTUD     │
            │    Int      │
            ├─────────────┤
  ──────────┤ CU      QU  ├──────────
    false ──┤ CD      QD  ├── false
    false ──┤ R       CV  ── 0
    false ──┤ LD          │
        3 ──┤ PV          │
            └─────────────┘
```

The "count up and down" counter (CTUD) counts or down by 1 on the 0 to 1 transition of the count up (CU) or count down (CD) inputs.

The timing diagram shows the operation of a CTUD counter with an unsigned integer count value (where PV = 4).

- If the value of parameter CV (current count value) is equal to or greater than the value of parameter PV (preset value), then the counter output parameter QU = 1.

- If the value of parameter CV is less than or equal to zero, then the counter output parameter QD = 1.

- If the value of parameter LD changes from 0 to 1, then the value at parameter PV is loaded to the counter as the new CV.

- If the value of the reset parameter R changes from 0 to 1, CV is reset to 0.

Box sorting application

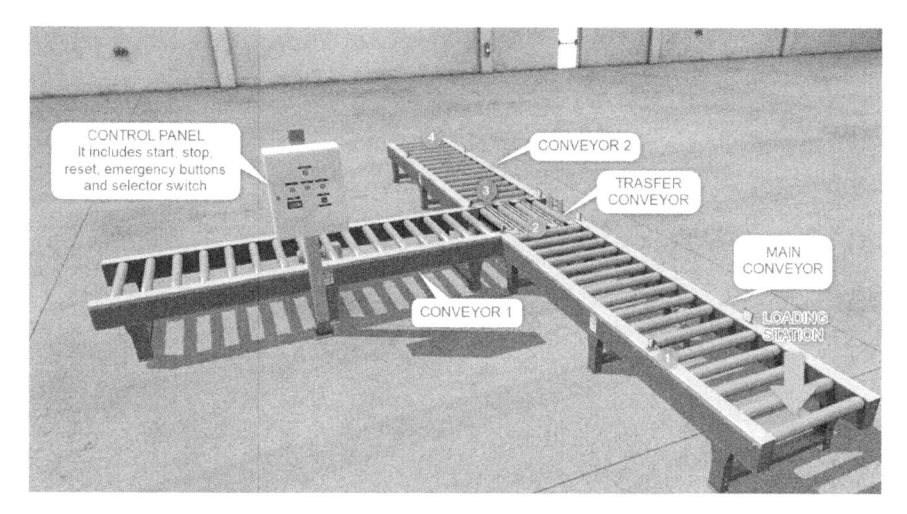

A box sorting application is shown in the above image. It consists of the main conveyor from where boxes are loaded and two unloading converters. Unloading conveyors carry the boxes in two different directions. Four reflex sensors have been used along with the reflectors as shown in the image.

When a reflex sensor detects the box on the main conveyor, the main conveyor will start after 1 second. Once a box approaches sensor 2, the main conveyor will stop.

Transfer conveyor will transfer 5 boxes to conveyor 1 first, and then it will transfer the remaining boxes to the conveyor 2. The counter gets reset when the stop button is pressed.

Input devices list

1. Start push button
2. Stop push button
3. Reflex sensor 1
4. Reflex sensor 2
5. Reflex sensor 3
6. Reflex sensor 4

Output devices list

1. Main conveyour motor
2. Conveyor 1 motor

3. Conveyor 2 motor
4. Transfer conveyor 1 motor
5. Transfer conveyor 2 motor

5. Math function

a. Add instruction

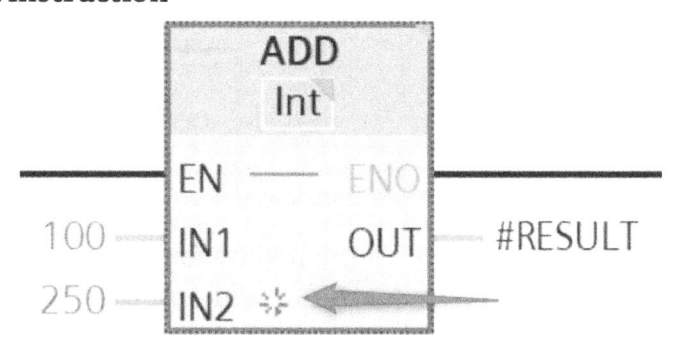

```
         ADD
         Int
    ─── EN    ENO ───
100 ─── IN1   OUT ─── #RESULT
250 ─── IN2 ◄───
```

ADD_I (Add integer) is activated by logic "1" at the Enabled (EN) input. IN1 and IN2 are added and the result can be scanned at OUT. If the result is outside the permissible range for an integer (16 bits), the OV bit and OS bit will be "1" and ENO is logic "0", so that other functions after this math box which are connected by the ENO are not executed.

b. Subtract instruction

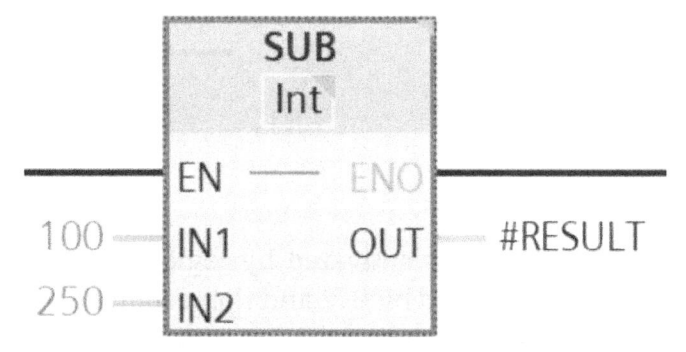

```
         SUB
         Int
    ─── EN    ENO ───
100 ─── IN1   OUT ─── #RESULT
250 ─── IN2
```

SUB_I (Subtract integer) is activated by logic "1" at the Enabled (EN) input. IN2 is subtracted from IN1 and the result can be scanned at OUT. If the result is outside the permissible range for an integer (16 bits), the OV bit and OS bit will be "1" and ENO is logic "0", so that other functions after this math box which are connected by the ENO are not executed.

c. Multiply instruction

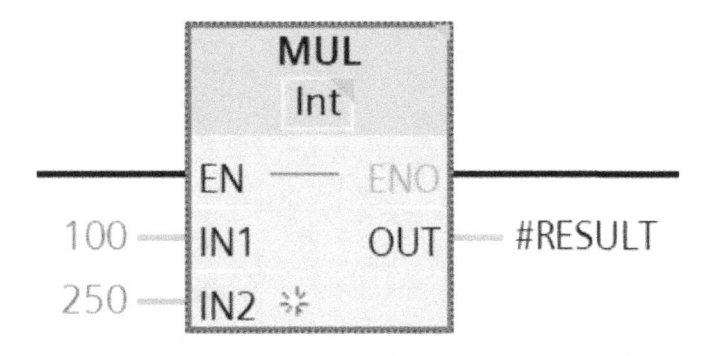

```
         MUL
         Int
      EN      ENO
100 — IN1     OUT — #RESULT
250 — IN2
```

MUL_I (multiply integer) is activated by logic "1" at the enable (EN) input. IN1 and IN2 are multiplied and the result can be scanned at OUT. If the result is outside the permissible range for an integer (16 bit), the OV bit and OS bit will be "1" and ENO is logic "0", so that other functions after this math box which are connected by the ENO are not executed.

d. Divide instruction

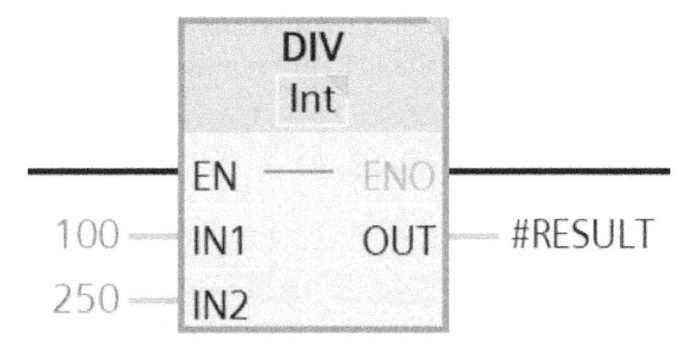

```
         DIV
         Int
      EN      ENO
100 — IN1     OUT — #RESULT
250 — IN2
```

DIV_I (divide integer) is activated by logic "1" at the enable (EN) input. IN1 is divided by IN2 and the result can be scanned at OUT. If the result is outside the permissible range for an integer (16 bit), the OV bit and OS bit will be "1" and ENO is logic "0", so that other functions after this math box which are connected by the ENO are not executed.

6. Move operation
a. MOVE

It copies a single data element from the source address specified by the IN parameter to the destination address specified by OUT parameter.

b. MOVE BLK

The MOVE BLK instruction has an additional count parameter. The count specifies how many data elements are copied. The number of bytes per element copied depends on the data type assigned to the IN and OUT parameter tag names in PLC tag table.

c. FILL BLK

Fill block instruction is used to fill a memory area with value of INT input, the destination area is filled beginning with the address specified at the output.

d. SWAP

SWAP instruction is used to change order of the bytes at input IN and query the result at output OUT.

7. **Comparator**

The compare instructions are used to compare a quantity with another quantity.

a. Equal/Not equal
b. Greater or equal/ Less or equal
c. Greater than/ Less than
d. In range
e. Out range

8. **Conversion operation**

a Scale_X- Scale

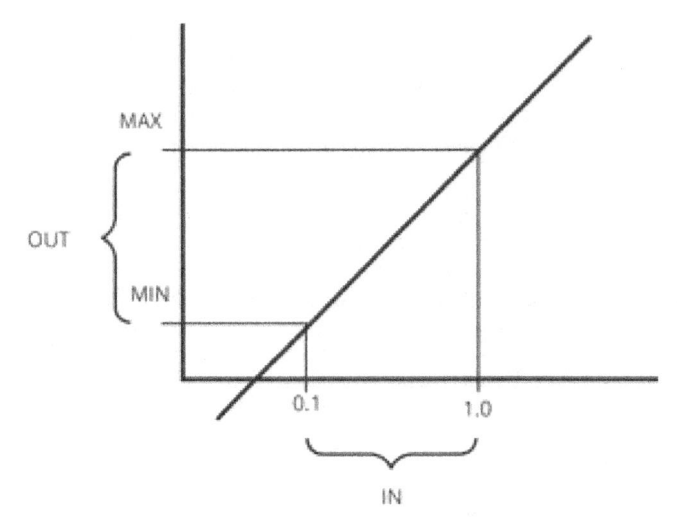

You can use the "Scale" instruction to scale the value at the VALUE input by mapping it to a specified value range. When the "Scale" instruction is executed, the floating-point value at the VALUE input is scaled to the value range that was defined by the MIN and MAX parameters. The result of the scaling is an integer, which is stored in the OUT output.

The "Scale" instruction works with the following equation:

$$OUT = [VALUE * (MAX - MIN)] + MIN$$

Enable output ENO has the signal state "0" if one of the following conditions applies:

- The EN enable input has the signal state "0".
- The value at the MIN input is greater than or equal to the value at the MAX input.
- The value of a specified floating-point number is outside the range of the normalized numbers according to IEEE-754.
- An overflow occurs.
- The value at the VALUE input is NaN (Not a Number = result of an invalid arithmetic operation).

b. Norm_X- Normalise

You can use the instruction "Normalize" to normalize the value of the tag at the VALUE input by mapping it to a linear scale. You can use the MIN and MAX parameters to define the limits of a value range that is applied to the scale. The result at the OUT output is calculated and stored as a floating-point number depending on the location of the value to be normalized within this value range.

If the value to be normalized is equal to the value at the MIN input, the OUT output has the value "0.0". If the value to be normalized is equal to the value at input MAX, output OUT returns the value "1.0".

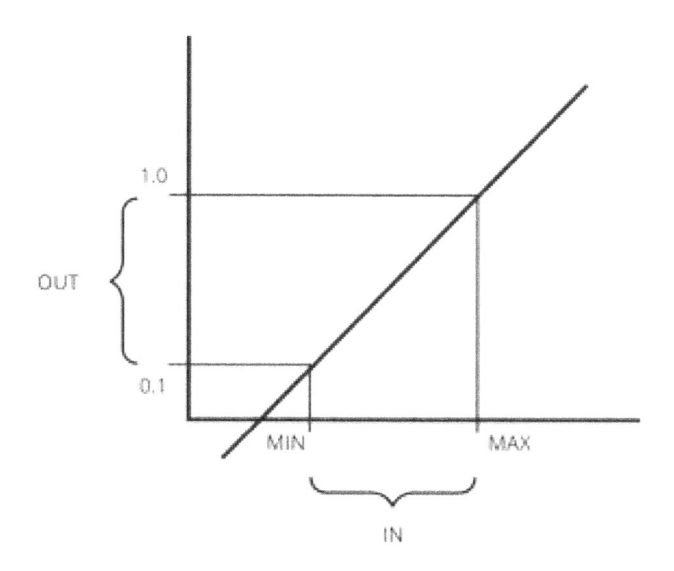

The "Normalize" instruction works with the following equation:

$$OUT = (VALUE - MIN) / (MAX - MIN)$$

Enable output ENO has the signal state "0" if one of the following conditions applies:
- The EN enable input has the signal state "0".
- The value at the MIN input is greater than or equal to the value at the MAX input.

The value of a specified floating-point number is outside the range of the normalized numbers according to IEEE-754.

The value at the VALUE input is NaN (result of an invalid arithmetic operation).

BCD_I (Convert BCD to integer) reads the contents of the IN parameter as a three digit, BCD coded number (+/-999) and converts it to an integer value (16 bit). The integer result is output by the parameter OUT. ENO always has the same signal state as EN.

I_BCD (Convert Integer to BCD) reads the content of the IN parameter as an integer value (16 bits) and converts it to a three digit BCD coded number (+/-999). The result is output by the parameter OUT. If the overflow occurred. ENO will be "0".

BCD_DI (Convert BCD to integer) reads the content of the IN parameter as a seven digit, BCD coded number (+/-9999999) and converts it to a double integer value (32 bits). The double integer result is output by the parameter OUT. ENO always has the same signal state as EN.

I_DINT (Convert Integer to Double Integer) reads the content of the IN parameter as an integer (16 bit) as converts it to a double integer (32 bit). The result is output by the parameter OUT. ENO always has the same signal state as EN.

DI_BCD (Convert Double Integer to BCD) reads the content of the IN parameter as a double integer (32bit) and converts it to a seven-digit BCD coded number (+/-9999999). The result is output by the parameter OUT. If an overflow occurred. ENO will be "0".

9. Program control operation

a. Switch Jump

MOVE (Assign a value) is activated by the enable EN input. The value specified at the IN input is copied to the address specified at the OUT output. ENO has the same logic state as EN.

MOVE can copy only BYTE, WORD, and DWORD data objects. User defined data types like arrays or structures have to be copied with the system function "BLKMOVE" (SFC 20).

b. Jump instruction

Jump functions as an absolute jump when there is no other ladder element between the left-hand power rail and the instruction.

c. Return instruction

Call is used to call a function or system function that has no passed parameters. A call is only executed if RLO is "1" at the CALL coil. If call is executed.

10. Word Logic operations

a. Decode

The decode instruction is used to set a bit in the output value that is specified by the input value.
The decode instruction reads the value at IN input and sets the bit in the output value whose bit position corresponds to the read value. The other bits in the output value are filled with zeros.

b. Encode

The encode instruction is used to read the bit number of the latest value and to send it to the OUT output.

c. Select

Depending upon the 'G Input' the select instruction selects one of the inputs IN0 or IN1 and copies its content to the OUT output.
All tags at all parameters must have same data type.

d. Mux

You can use this instruction to copy the content of a selected input to output OUT. The number of selectable inputs of the instruction box can be expanded.

e. Demux

You can use this instruction to copy the content of the input IN to a selected output. The number of selected outputs can be expanded in the instruction.

11. Left/ right shift

This instruction is used to shift the content of the operand at the input IN bit by bit to the right and query the result at OUT output. You can use the N parameter to specify the number of the bit positions by which the specified value to be shifted.

12. Left/Right rotate

It is used to rotate the content of the operand at the input IN bit by bit to right and query the result at the OUT. You can use N parameter to specify the number of bit by which the specified value to be rotated.

D. Troubleshooting

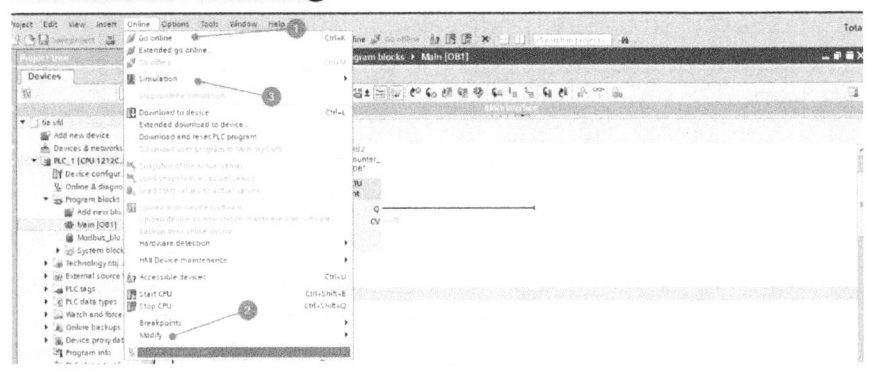

Go online- - In this mode you can observe the IO devices as well as make some changes in program without disturbing machine's operation.

Modify- To make changes in logic you just have to modify the value from 0 to 1 or 1 to 0. After making changes in program, click on convert option and then select online program change.

If a machine gets broken down during operation and you observe its one of the outputs is not working. You can troubleshoot by online monitoring of a program. Online monitoring takes less time as compare to physical tracing of wires.

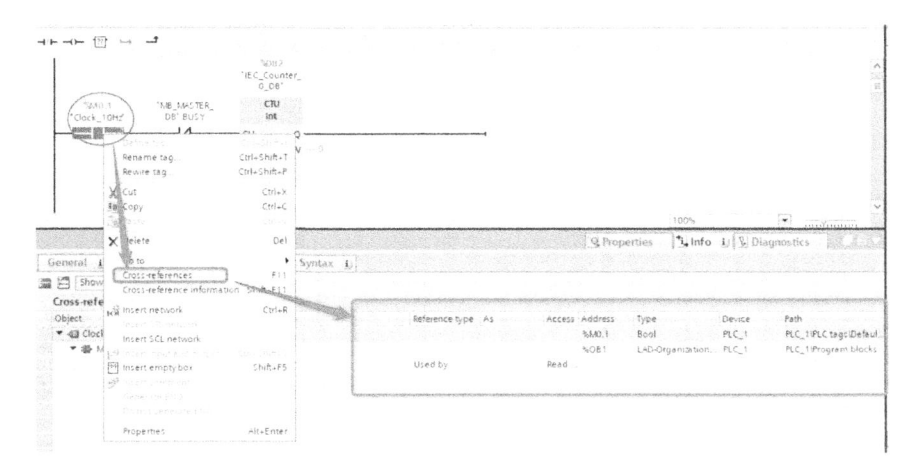

1. Select the bit you want to trace, and right click on it, then select cross reference.

2. A chart will be displayed on your screen, it will show the use of that bit in different networks. From that chart you can easily monitor the bit.

E. KTP 700 HMI Programming

In this section we are going to cover the HMI development part. I have chosen siemens KTP 700 HMI for demonstration. You do not need another software for HMI programming.

1. Adding an HMI device

New HMI devices can be added in both the Portal view and the Project view. More than anything else, attention must be paid to the device data such as article (order) number and version number.

In order to get access rights to a CPU that is password-protected, the HMI device must log on to the CPU with a password when Runtime is started. This password must be specified in the Connections configuration of the HMI device (see picture bottom right.

In order to get access rights to a CPU that is password-protected, the HMI device must log on to the CPU with a password when Runtime is started. This password must be specified in the Connections configuration of the HMI device (see picture bottom right).

1. Configure the connection to the PLC.
2. Select the background color for the template and the elements for the header.
3. Disable the system screens. These are not necessary for the example project.
4. Enable the lower button area and insert the "Exit" button. Runtime can be terminated using this button.
5. Click the "Save project" button on the toolbar to save the project.

Creating and configuring graphic objects
1. Machine ON/OFF button

The following steps show you how to create the "Machine ON/OFF" button and how to connect it to the PLC tag "ON_OFF_Switch" using an external HMI tag. You can use this approach to modify the process values of the PLC tag by means of the HMI screen.

You use an external HMI tag to access a PLC address. This allows you, for example, to input a process value via an HMI device or to directly modify process values of the control program via a button. Addressing takes place via the PLC tag table in the PLC linked to the HMI device. The PLC tag is linked to the HMI tag with the symbolic name. This means you do not have to adapt the HMI device when you change the address in the PLC tag table.

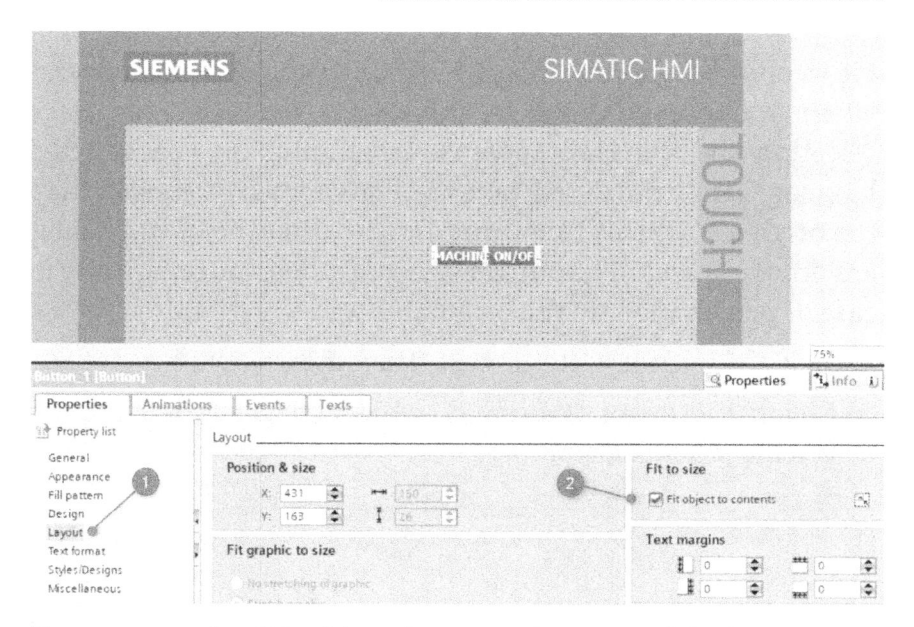

To connect the "Machine ON/OFF" button with the PLC tag "ON_OFF_Switch", follow these steps:

a. In the Inspector window, select the "Fit object to contents" option, to automatically adjust the button size to the text length.

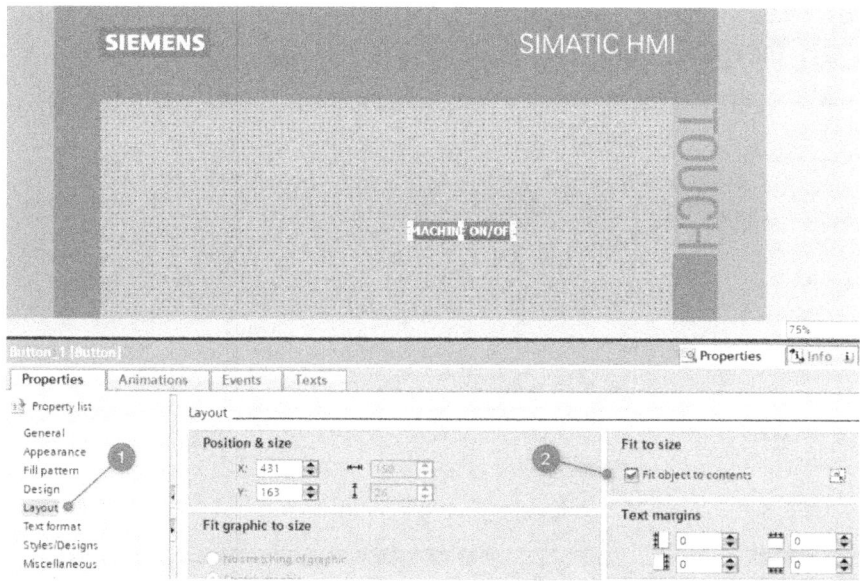

Fit to object function-You can use this function in particular when you are working in future projects with language selection for the HMI screens. Depending on the selected language, the translated text can be shorter or longer than the original. Use this function to ensure that the button labels are not truncated. The button size will adjust automatically in case of text changes in the original.

b. Label the button with the text "Machine ON/OFF".

c. Assign the event "Pressing", which triggers the function "InvertBit", to the button.

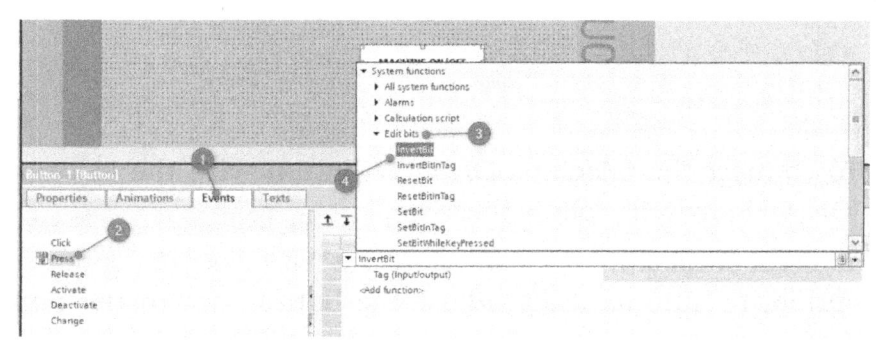

d. Link the "InvertBit" function to the PLC tag "ON_OFF_Switch".

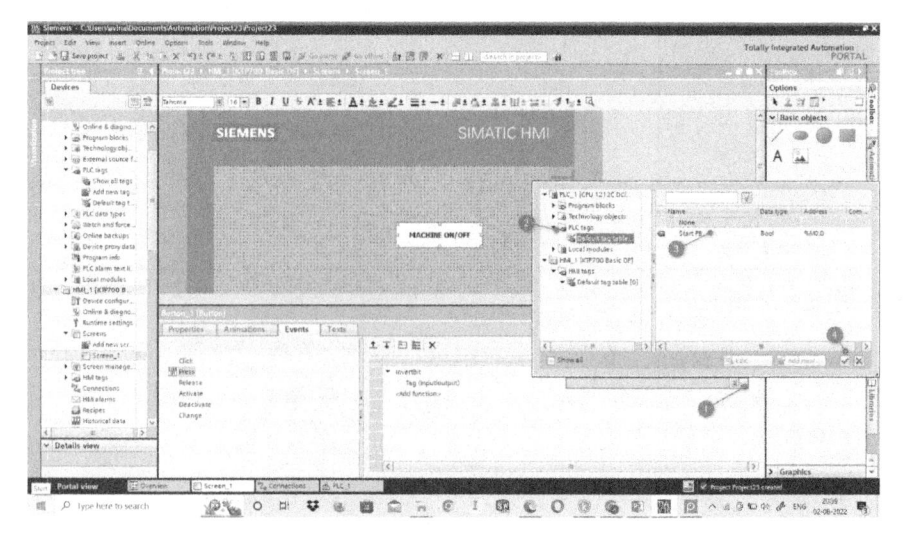

e. Assign PLC tag to button

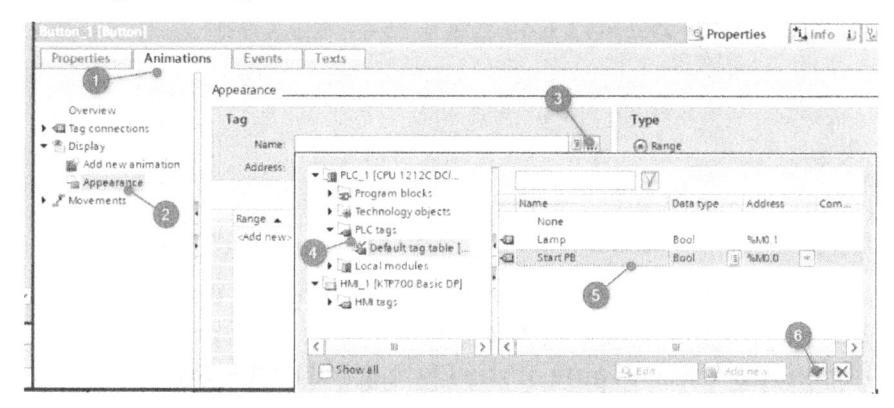

1. Go to animation.
2. Select appearance.
3. Click on three dots.
4. Choose a PLC tag table.
5. Select a PLC tag.
6. Click on OK.

f. Button color in ON/OFF condition

You can select the color of button in on and off condition.

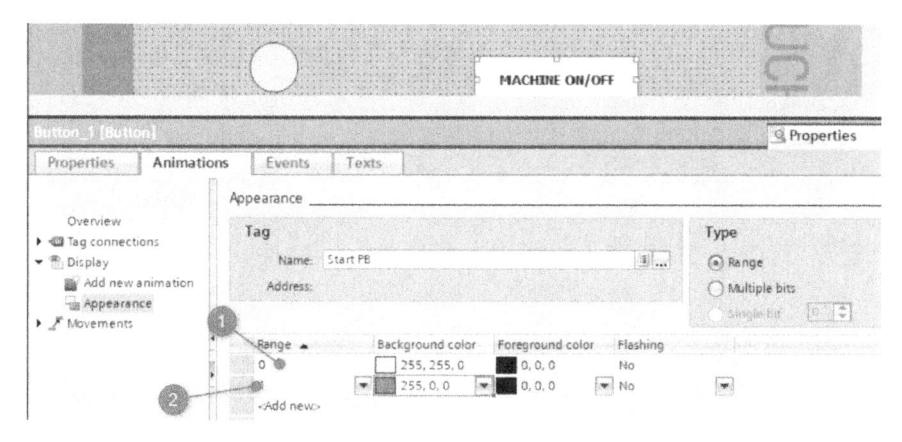

1. Yellow color of LED in off condition.
2. Red color of LED in on condition.

2. Graphic objects "LEDs"

The LED shows the condition of any output device.

1. Create a LED graphic object.

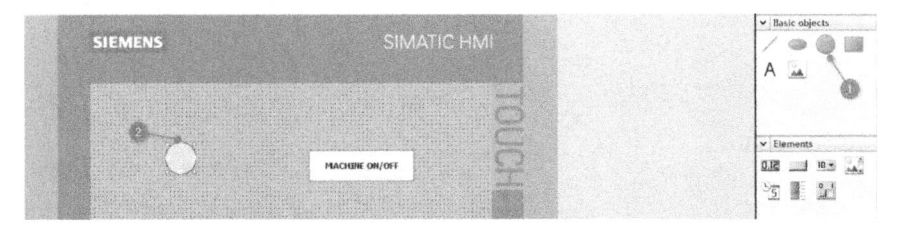

The following steps show you how to set up two status LEDs (red/green) using the circle object and animate them depending on the value of the PLC tag "ON_OFF_Switch".

1. Right click on LED to open property
2. Appearance- You can change the LED color from here
3. Select PLC tag

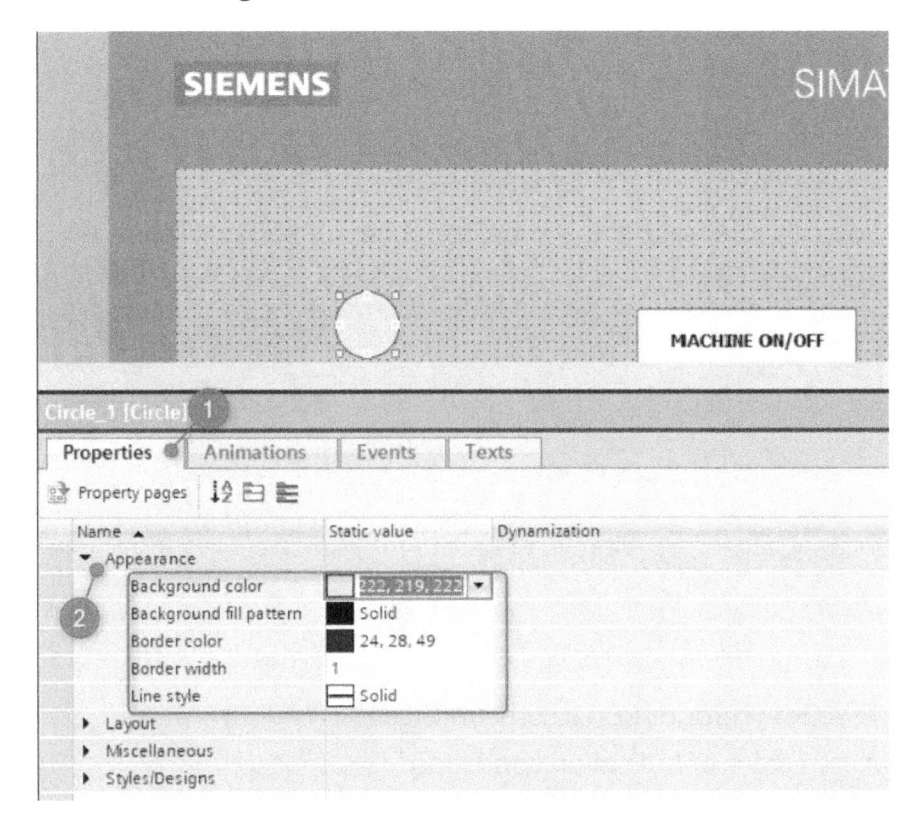

In this screen you can assign a PLC tag to LED lamp.

1. Go to animation
2. Select appearance
3. Click on three dots
4. Choose a PLC tag table & Select a PLC tag
5. Click on OK

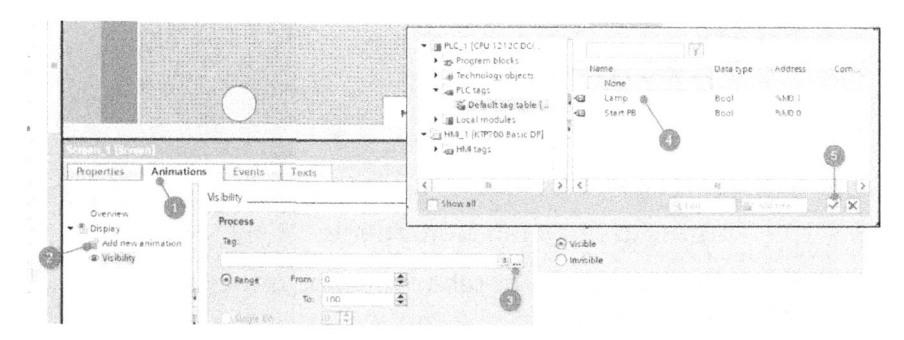

3. LED color in ON/OFF condition

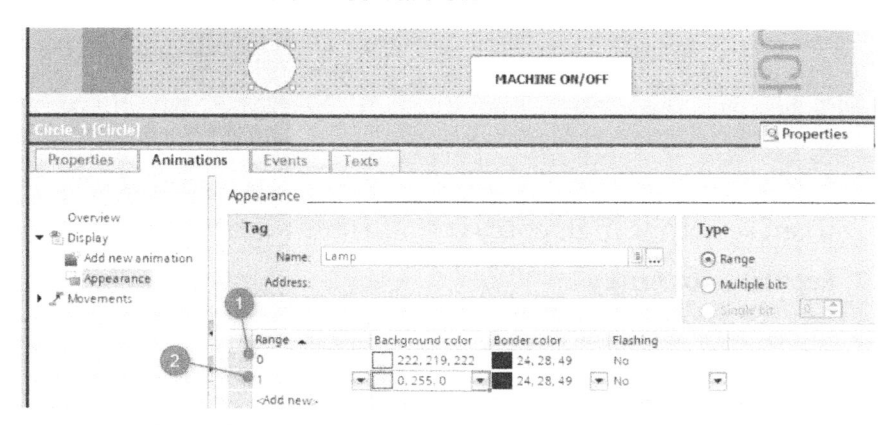

1. Grey color of LED in off condition
2. Green color of LED in on condition

Numeric value entry from HMI

HMI comes with a facility to change the value of parameters such as Frequency, time, etc. To accept the value from outside we need to use numeric input tab in HMI screen during development.

1. Create a numeric entry tab

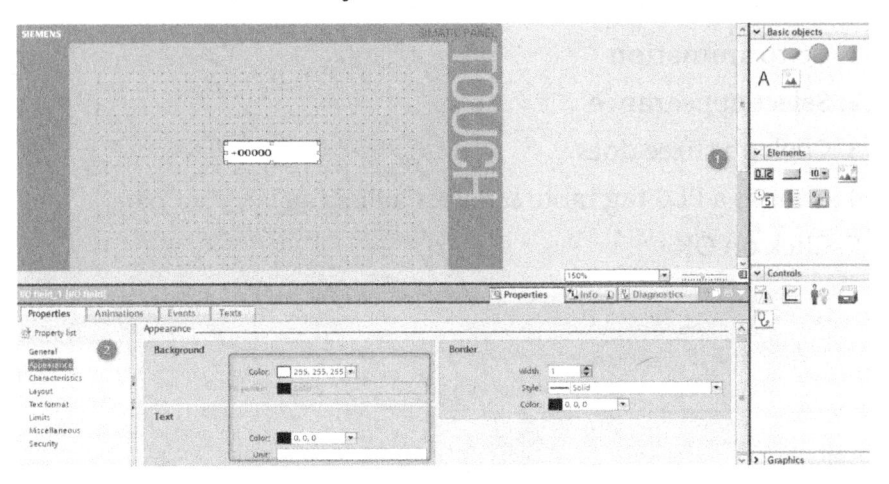

1. Select a numeric entry tab from menu
2. Change the appearance of tab from appearance section

2. Entry limit

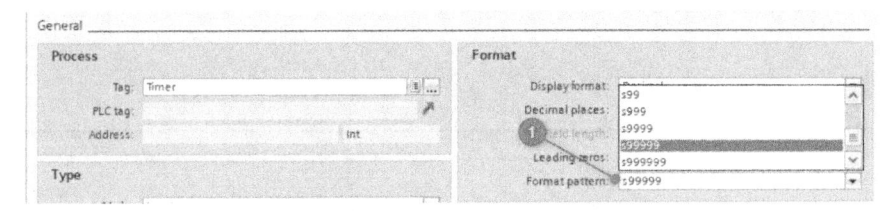

You can select the digit entry from format section.

3. Security for entry

LEARN EVERYTHING ABOUT PLC PROGRAMMING

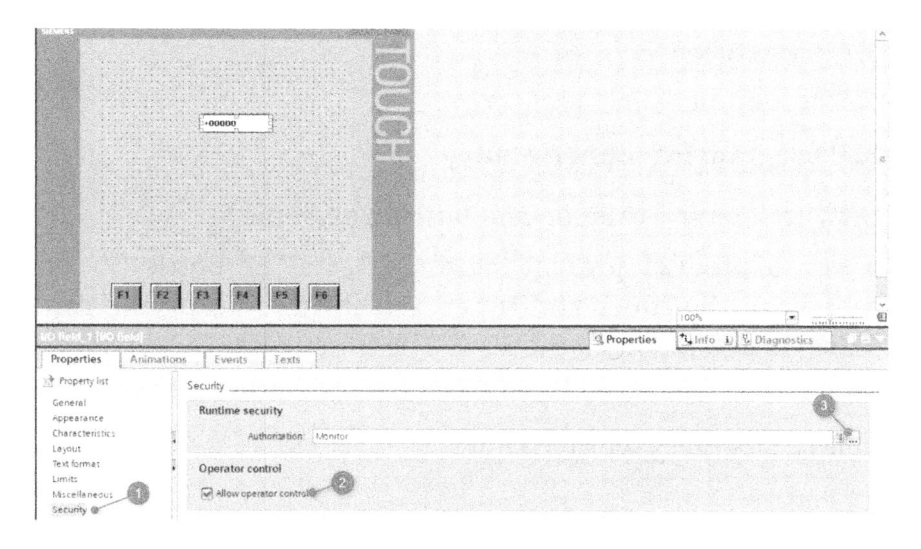

1. Right click on the tab and select security
2. Check box for operator control
3. Select authorization

4. Assign a PLC tag

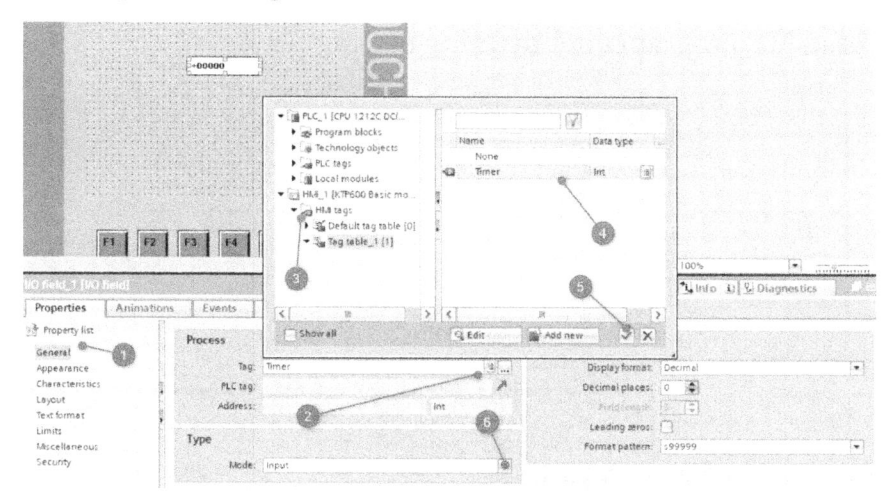

1. Right click on the tab and select general
2. Click on three dots
3. Select HMI tag table
4. Select a HMI tag

197

5. Click on OK

6. Select input

PLC program for numeric entry

Let's understand this concept with an example.

When a product present sensor detects the bottle, it will switch on the timer and timer will start counting the time which is entered from the HMI. The timer value is first stored in memory word MW20, then it is moved to Timer.PT.

Define a memory word to timer

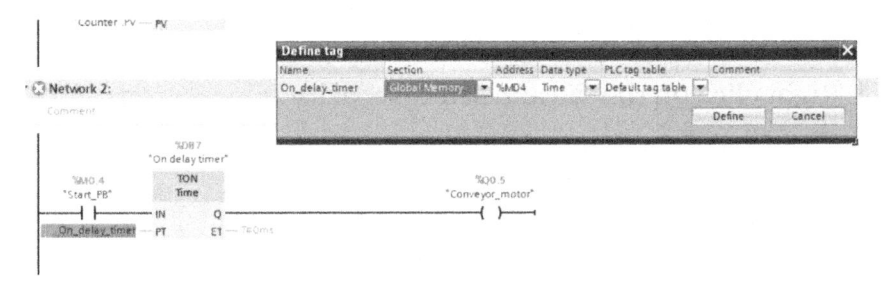

Read PLC program after defining PLC tag

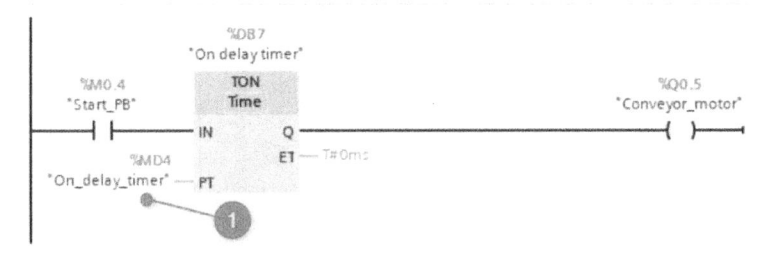

HMI and PLC connection

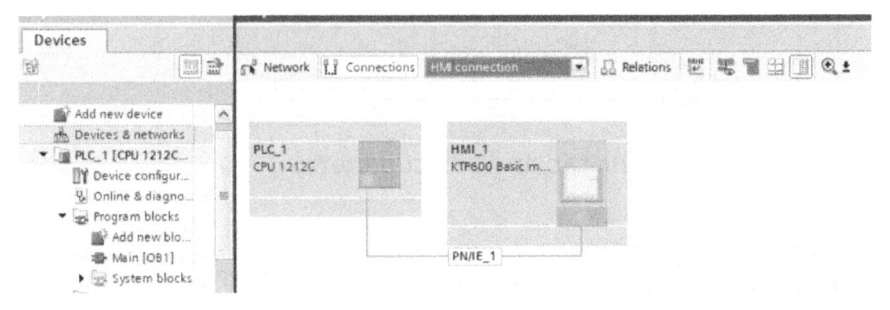

Synchronization with PLC tags

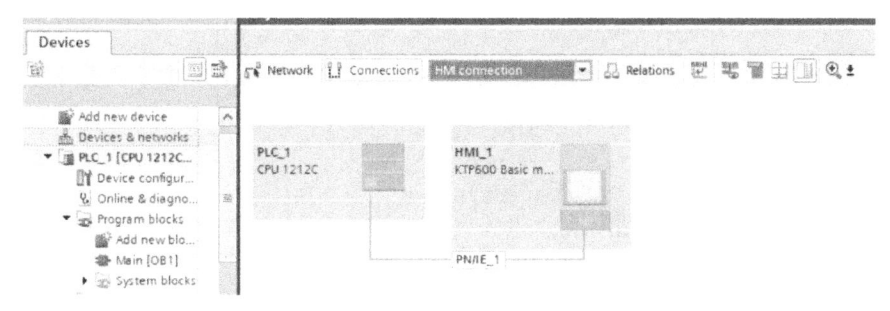

Allocate HMI tags

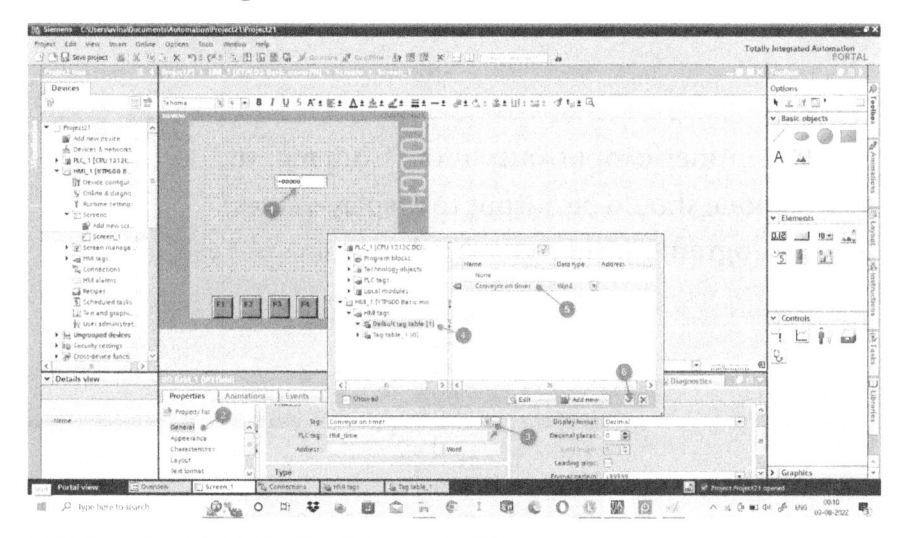

Numeric value display on HMI

The Numeric value display function is used to display certain values on HMI screen. You can use this function to display production report, value of timer, counter, etc.

1. Create a numeric display tab

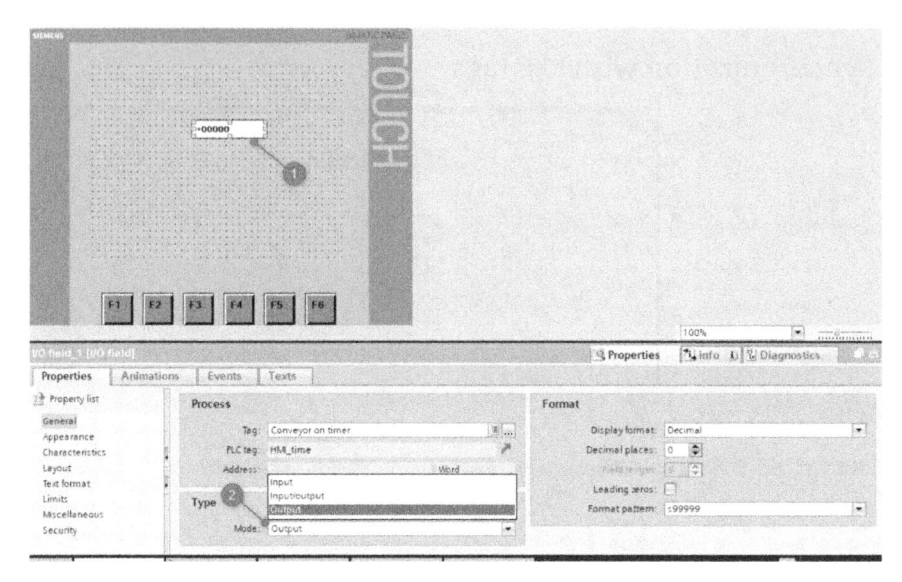

1. Select numeric input/output tab from menu.
2. The mode should be output to display values.

2. PLC program for numeric display

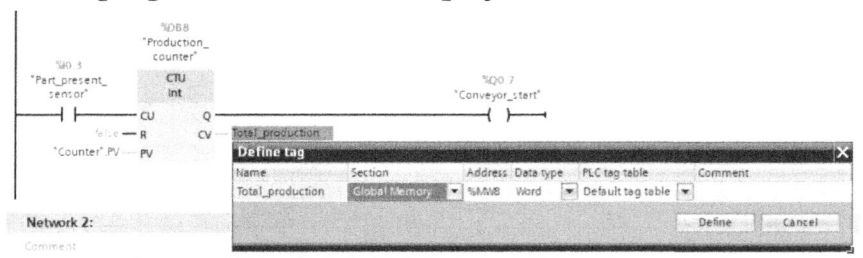

After defining PLC tags, you can see the rung as shown below.

3. Allocate HMI tag to I/O field

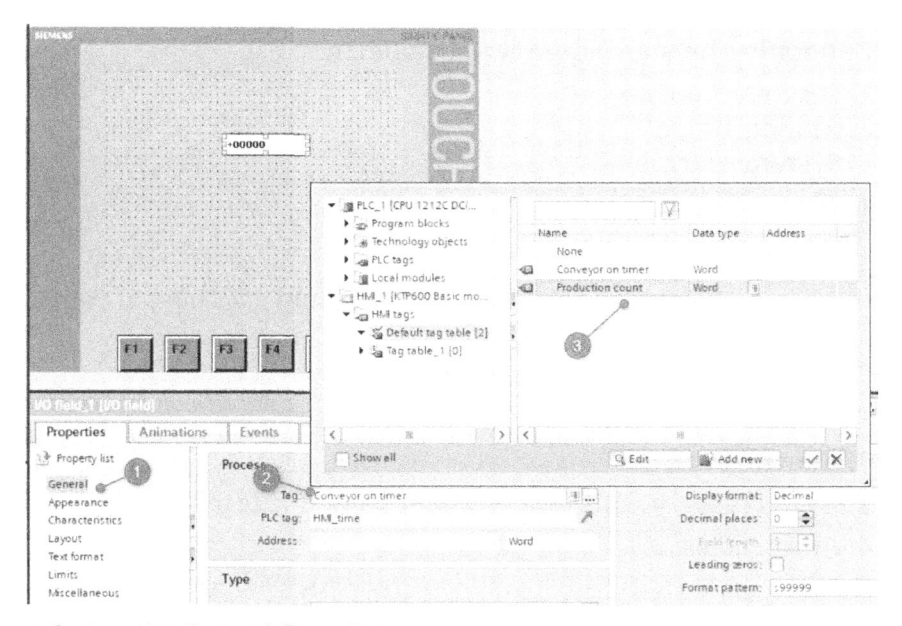

Alarm and notification

The alarms are the most important part of HMI development. In this section we are going to cover the creation of alarms using TIA portal.

Create Alarm screen

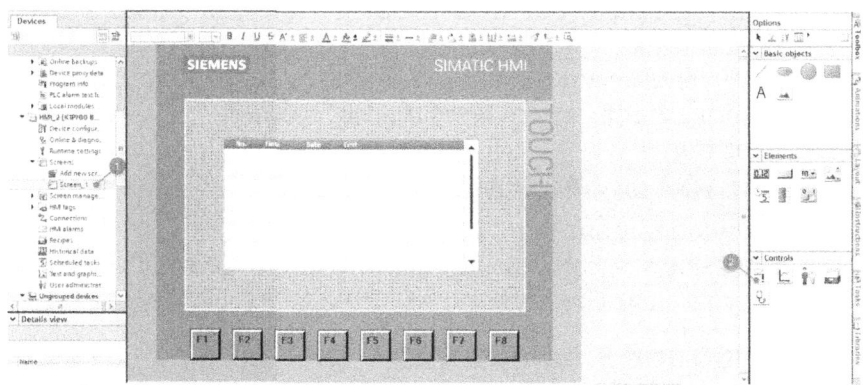

Creating a screen for Alarms is very simple method.

1. Create a seperate screen for Alarm.
2. Alarm view- Select an alarm view tab from menu, a table gets drawn infront of you. This table shows the alarms with time and date.

Generally, there are three types of alarms:
1. Discrete Alarm
2. Analog alarm
3. System Alarms

<u>1. Discrete Alarm</u>

1. Alarm ID

It shows the number of alarms assigned to the particular alarm.

2. Name

You can rename it for your understanding

3. Alarm Text

The HMI displays this text on the screen, if this tag is triggered.

4. Alarm class

Alarm classes							
Display name	Name	State machine	Log	Backgro...	Backgro...	Backgro...	Backgro...
!	Errors	Alarm with single-mode	<No log>	255...	255...	255...	255...
!	Warnings	Alarm without acknowle...	<No log>	255...	255...	255...	255...
S	System	Alarm without acknowle...	<No log>	255...	255...	255...	255...
A	Acknowledgement	Alarm with single-mode	<No log>	255...	255...	255...	255...
NA	No Acknowledgement	Alarm without acknowle...	<No log>	255...	255...	255...	255...
<Add new>							

Errors- These types of alarms require acknowledgement to get cleared. For example an alarm gets displayed on HMI if a emergency pressed, but it will not get cleared untill a reset button is pressed.

Warning- These types of alarms do not need any type of acknowledgement. It gets reset if a tag is false.

5. Trigger tag

Assign an HMI tag by following this procedure.

6. Acknowledgement tag

You can assign a tab for tag acknowledgement. This tag is necessary to clear the fault.

F. S7 1200 and Sinamics V20 VFD

In this particular section, we have chosen a Sinamics V20 VFD which is controlled by S7 1200 PLC with Modbus communication. In this section we will cover:

The hardware configuration consists of:

1. PLC- S7 1200

2. Modbus RTU CM1241 module
3. HMI- KTP 700
4. Sinamics V20 VFD

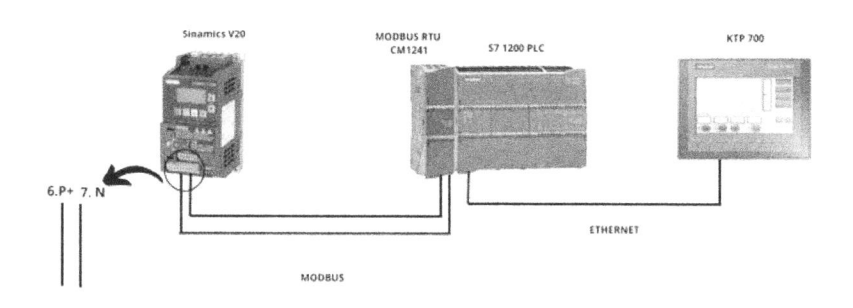

KTP 700 HMI screen

This is a KTP 700 HMI developed using TIA portal V16 software. By using this HMI screen, we can control and monitor the VFD.

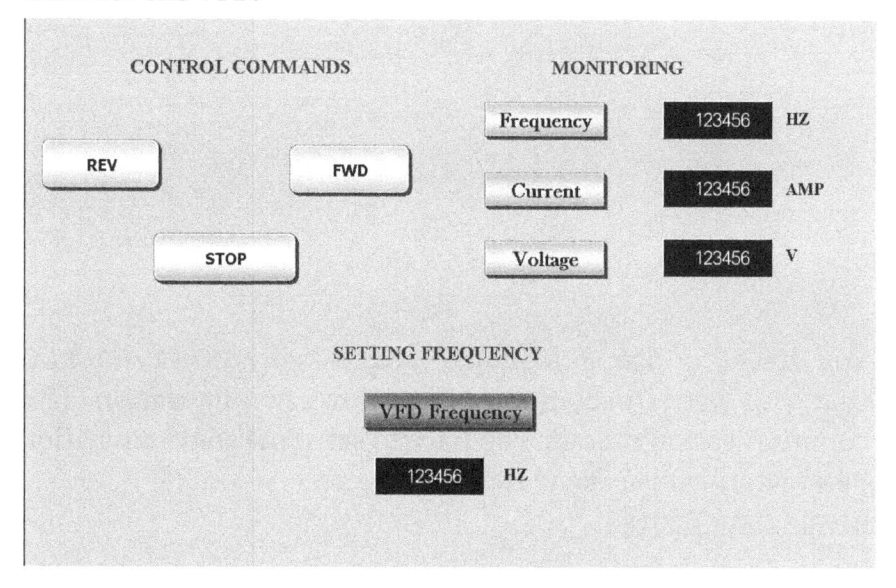

Sinamics V20 VFD Modbus communication parameters
This is the list of parameters which needed to be set in VFD for Modbus communication.

Parameter	Description	Setting	Remark
P0003	User access level	3	Expert: For expert use only
P700	Selection of command source	5	USS/MODBUS on RS485
P2010	USS/MODBUS baud rate	6	9600 bps
P2014	USS/MODBUS telegram off time (ms)	3000	3000ms
P2021	Modbus address	1	Slave no 1
P2022	Modbus reply timeout	1000	1000ms
P2023	RS485 protocol selection	2	Modbus
P2034	Modbus parity on RS485	2	Even
P2035	Modbus stop bits on RS485	1	Stop bit = 1

S7 1200 CM1241 Parameters Settings

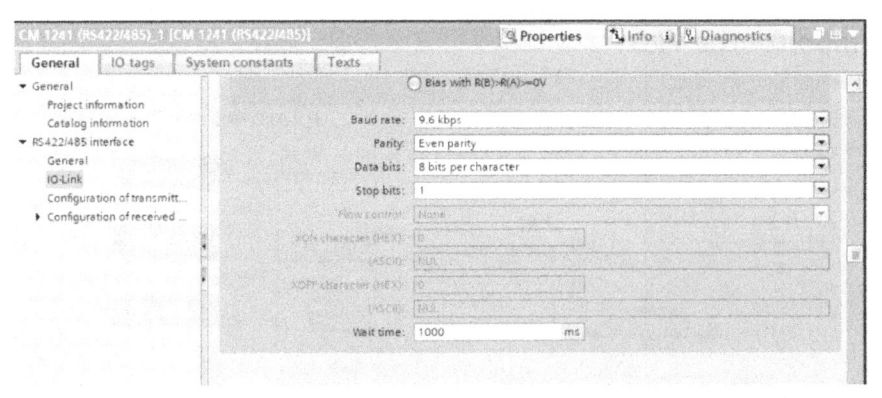

You have to use a Modbus module with S7 1200 PLC, that you need to configure first. After configuration of a communication module you have to set some communication parameters such as baud rate, parity, etc.

Create Data Types

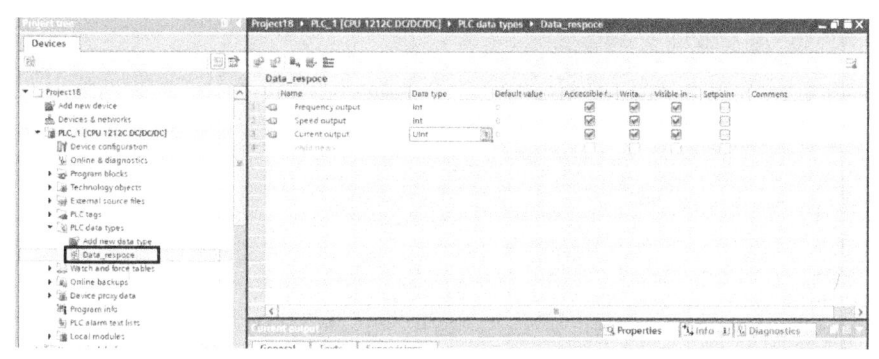

The next step is to create a data response block, that will collect information like frequency, speed, current, etc. from VFD.

Create Modbus Block

You have a create a modbus communication block for VFD.

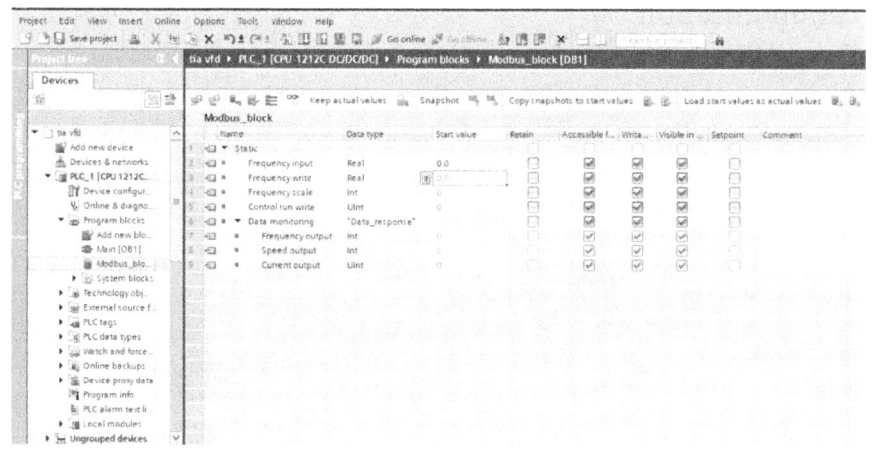

Automatic Read/Write Distribution

Communication Configuration

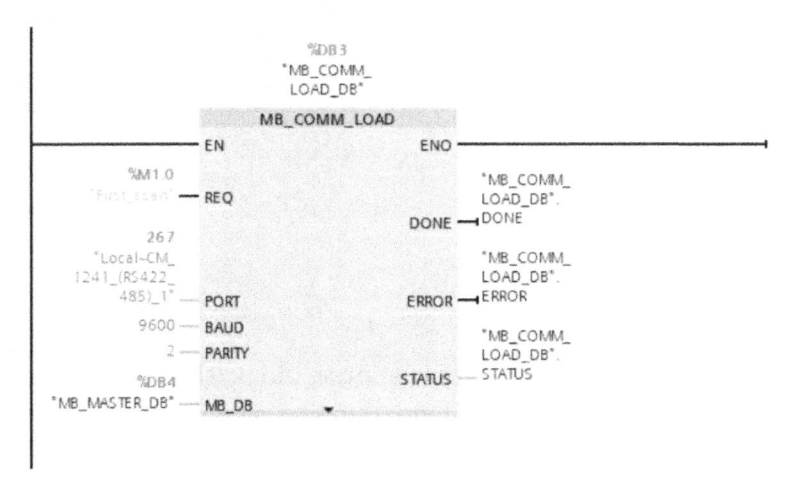

You need to configure a data block for Modbus communication.

Frequency Setting Command

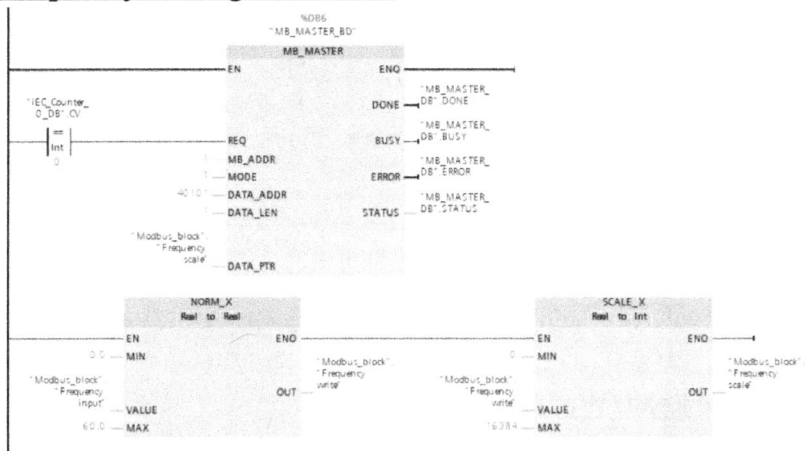

Motor Control Command

VFD Forward run- 1151

VFD Reverse run- 3199

VFD Stop- 1150

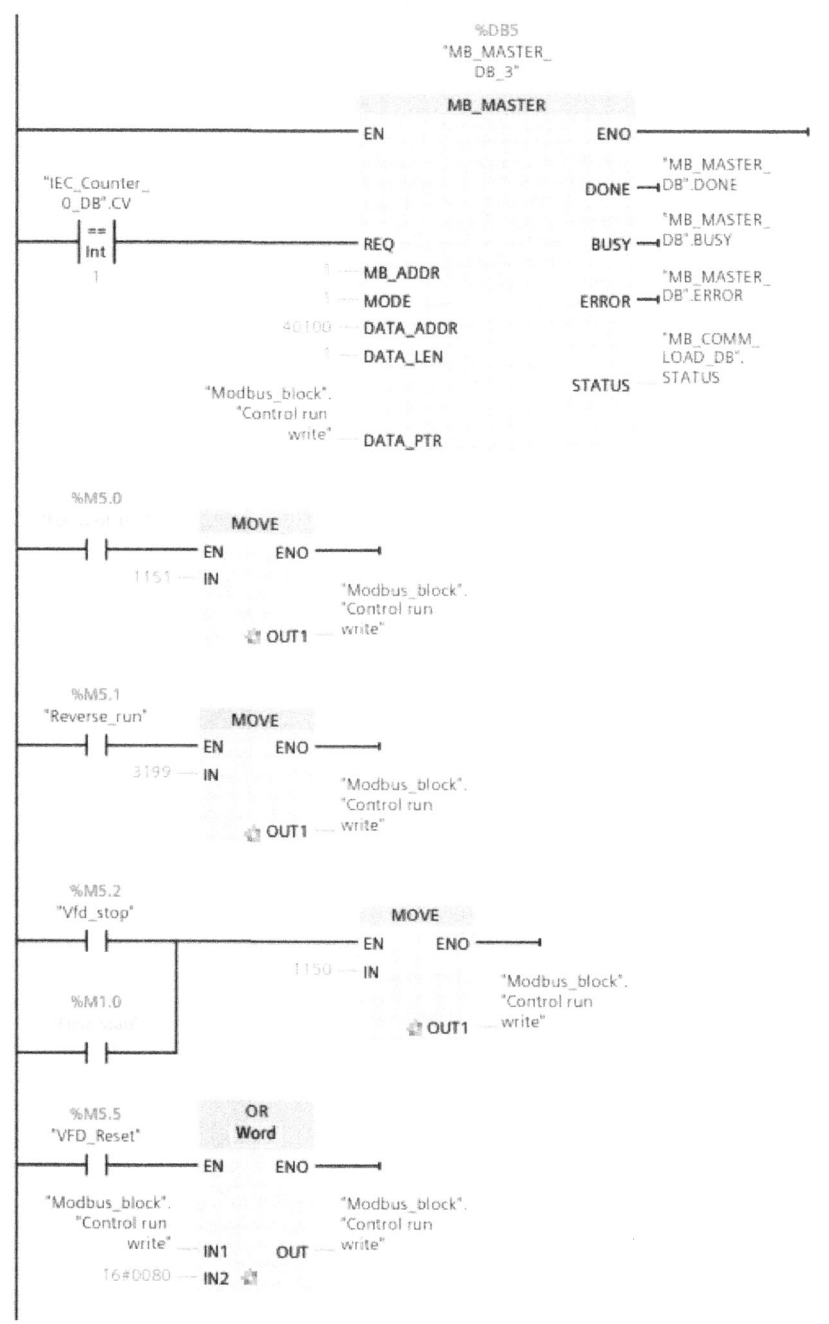

Read Data Response

The PLC reads the value from VFD register and then displays them on HMI. Refer the sinamics V20 for more information.

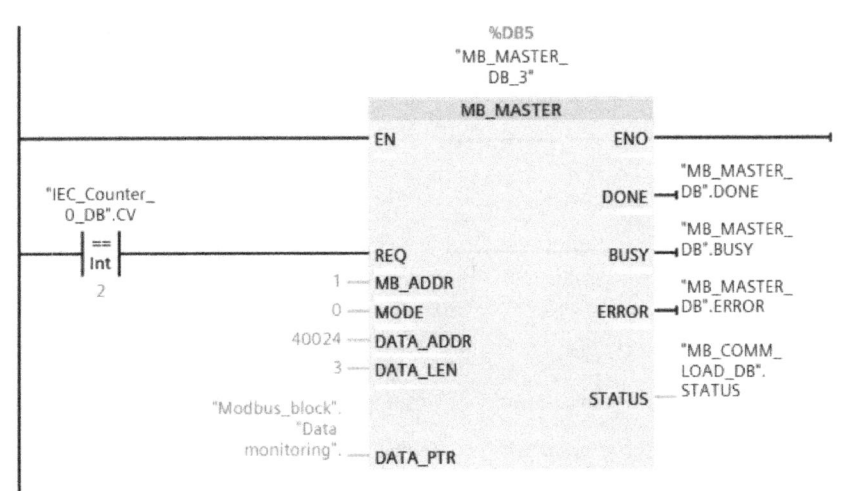

```
                              %DB5
                           "MB_MASTER_
                              DB_3"
                          ┌──────────────┐
                          │   MB_MASTER  │
                        ──┤ EN        ENO├──────────────────
                          │              │          "MB_MASTER_
   "IEC_Counter_          │          DONE├──DB".DONE
     0_DB".CV             │              │          "MB_MASTER_
      ──┤==├──────────────┤ REQ      BUSY├──DB".BUSY
         │Int│           1┤ MB_ADDR      │          "MB_MASTER_
          2             0─┤ MODE    ERROR├──DB".ERROR
                    40024─┤ DATA_ADDR    │          "MB_COMM_
                        3─┤ DATA_LEN     │          LOAD_DB".
                          │        STATUS├── STATUS
         "Modbus_block".  │              │
              "Data       │              │
           monitoring".──┤ DATA_PTR     │
                          └──────────────┘
```

Register No.		Description	Access	Unit	Range or On/Off text	Scaling factor
Inverter	MODBUS					
23	40024	Frequency output	R	Hz	-327.68 - 327.67	100
24	40025	Speed output	R	rpm	-16250 - 16250	1
25	40026	Current output	R	A	0 - 163.83	100

CASE STUDY 3

Pneumatic hammering machine

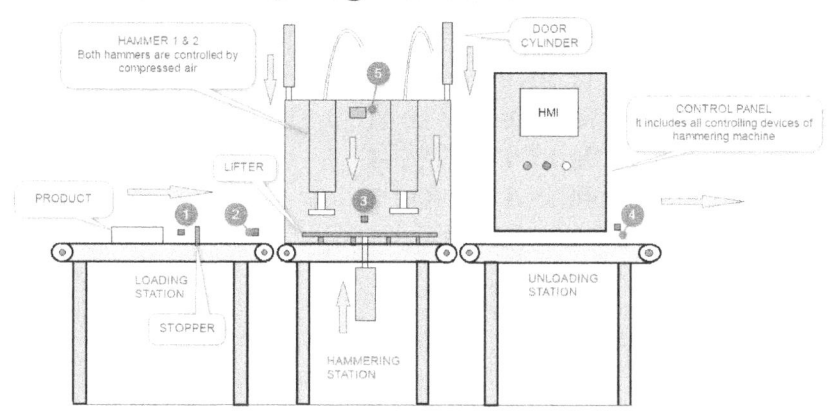

The pneumatic hammering machine consists of pneumatic hammers that are used to apply continuous hammering pressure on the casting part to remove the sand from the casted part.

Hardware configuration

PLC Controller- S7 1200
HMI- KTP 700
PLC & HMI communication- Ethernet
Motor control- Hardwire (via contactor)

Input devices

Start PB (M0.0)
Stop PB (M0.1)
Emergency PB (I0.0)
1. Product present sensor1 (I0.1)
2. Product present sensor2 (I0.2)
3. Product present sensor3 (I0.3)
4. Product present sensor4 (I0.4)
5. Door limit switch (I0.5)
Door cylinder down reed switch (I0.6)
Door cylinder up reed switch (I0.7)
Lifter up reed switch (I1.0)

Auto/man selector switch (M1.2)

Output devices

Loading station conveyor (Q0.0)
Hammering station conveyor (Q0.1)
Unloading station conveyor (Q0.2)
Pneumatic hammer solenoid coil (Q0.3)
Door cylinder1 solenoid coil (Q0.4)
Door cylinder2 solenoid coil (Q0.5)
Lifter cylinder up solenoid coil (Q0.6)
Stopper cylinder coil (Q0.7)
Red lamp (Q1.0)
Green lamp (Q1.1)
Amber lamp (Q1.2)
Lifter cylinder down solenoid coil (Q1.3)
Hooter (Q1.4)

Operating procedure

A product is loaded manually at the loading station. If the machine is in auto mode and there is no fault (emergency pressed) then the machine should start when start PB is pressed.

If a product is present inside the cabin (sensor 3), then the door1 will remain closed and the stopper will pop up and it will stop the product. Door 1 will open only when a product is not present at the hammering station.

Once the product reaches sensor 3, the hammering station conveyor will stop after 1 second and both doors will get closed. A lifter will lift the product and start the hammer after 2 seconds. The hammering operation will continue as per set the time from HMI. After completion of the hammering operation, a lifter will come down, both doors will go up and restart the hammering conveyor.

Keep unloading station conveyor in running mode. Sensor 4 will keep counting the production.

HMI screen

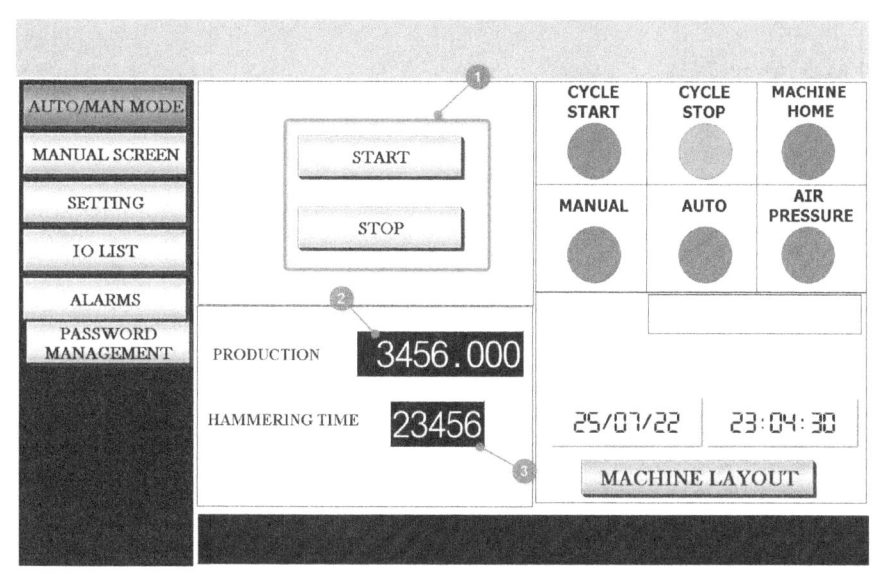

[HMI Screen showing: AUTO/MAN MODE, MANUAL SCREEN, SETTING, IO LIST, ALARMS, PASSWORD MANAGEMENT buttons on left; START, STOP buttons in center; CYCLE START, CYCLE STOP, MACHINE HOME, MANUAL, AUTO, AIR PRESSURE indicators on right; PRODUCTION 3456.000, HAMMERING TIME 23456, date 25/07/22, time 23:04:30, MACHINE LAYOUT button]

1. Control commands- To control the machine from HMI, you need to assign memory bits to the control tabs. I have assigned memory bits to the all tabs wherever I required machine to be controlled from HMI.

2. Numeric display- It shows the timer's value on HMI, which is stored in memory word. During HMI development you have to assign memory word MW to the display tabs.

3. Numeric entry- A PLC accepts the value from HMI, for that you need to use a numeric entry tab and have to assign a memory word.

Configuration of devices

PLC tags

I have created a list of PLC tags which includes input and output devices.

Default tag table

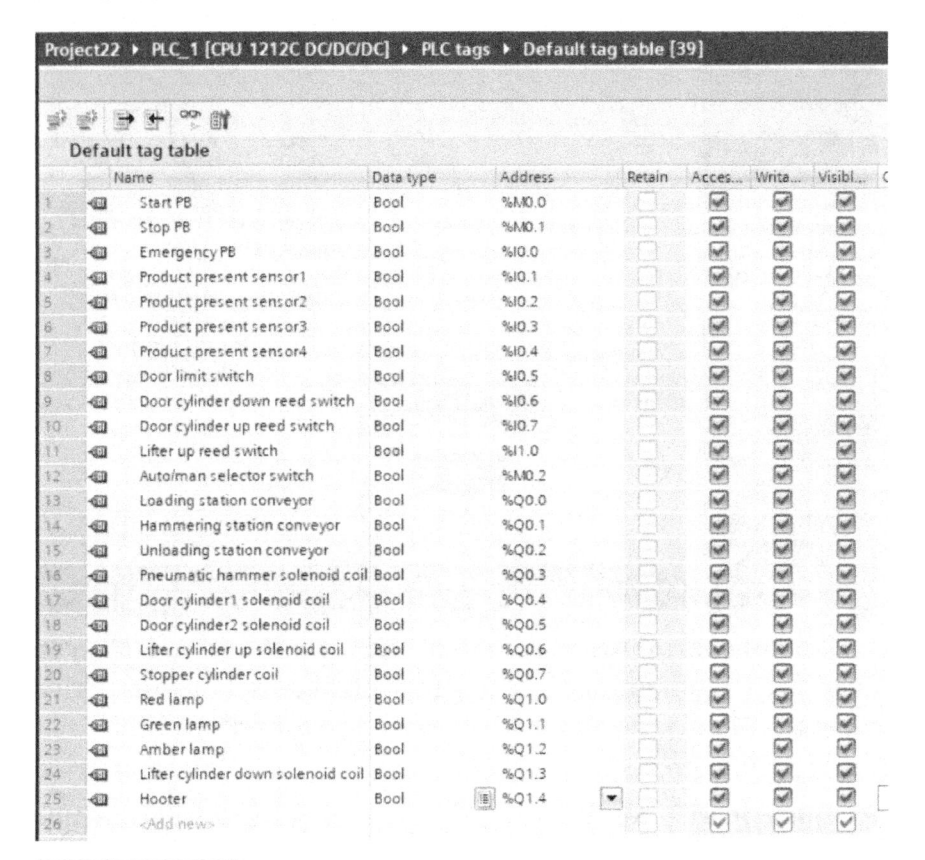

#	Name	Data type	Address
1	Start PB	Bool	%M0.0
2	Stop PB	Bool	%M0.1
3	Emergency PB	Bool	%I0.0
4	Product present sensor1	Bool	%I0.1
5	Product present sensor2	Bool	%I0.2
6	Product present sensor3	Bool	%I0.3
7	Product present sensor4	Bool	%I0.4
8	Door limit switch	Bool	%I0.5
9	Door cylinder down reed switch	Bool	%I0.6
10	Door cylinder up reed switch	Bool	%I0.7
11	Lifter up reed switch	Bool	%I1.0
12	Auto/man selector switch	Bool	%M0.2
13	Loading station conveyor	Bool	%Q0.0
14	Hammering station conveyor	Bool	%Q0.1
15	Unloading station conveyor	Bool	%Q0.2
16	Pneumatic hammer solenoid coil	Bool	%Q0.3
17	Door cylinder1 solenoid coil	Bool	%Q0.4
18	Door cylinder2 solenoid coil	Bool	%Q0.5
19	Lifter cylinder up solenoid coil	Bool	%Q0.6
20	Stopper cylinder coil	Bool	%Q0.7
21	Red lamp	Bool	%Q1.0
22	Green lamp	Bool	%Q1.1
23	Amber lamp	Bool	%Q1.2
24	Lifter cylinder down solenoid coil	Bool	%Q1.3
25	Hooter	Bool	%Q1.4
26	<Add new>		

Ladder Logic

Network 1:

```
  %M0.0                                      %M0.3
 "Start PB"                                "Latch bit"
───┤ ├─────────────────────────────────────────( S )───

  %M0.1                                      %M0.3
 "Stop PB"                                 "Latch bit"
───┤ ├─────────────────────────────────────────( R )───
```

When a start PB is pressed by an operator, a memory bit M0.0 gets set and gets reset when stop PB is pressed.

Network 2:

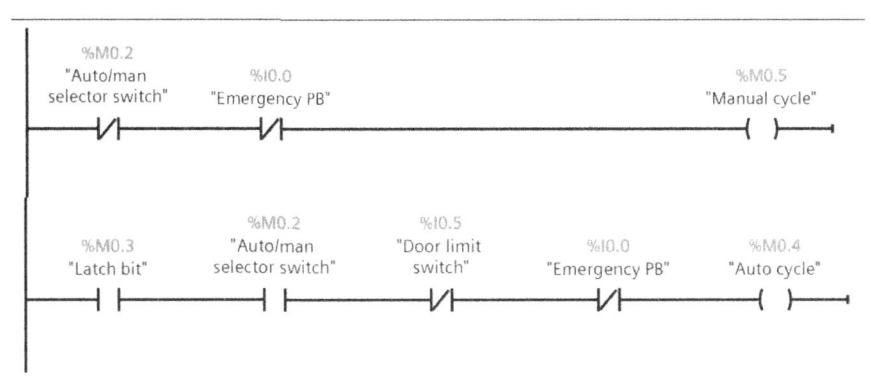

```
      %M0.2
    "Auto/man           %I0.0                                                              %M0.5
  selector switch"   "Emergency PB"                                                     "Manual cycle"
────┤/├──────────────┤/├───────────────────────────────────────────────────────────────────( )────

                        %M0.2              %I0.5
     %M0.3            "Auto/man         "Door limit         %I0.0              %M0.4
   "Latch bit"      selector switch"      switch"       "Emergency PB"       "Auto cycle"
────┤ ├──────────────┤ ├───────────────┤/├────────────────┤/├──────────────────( )────
```

A machine will start in auto mode only when all the above bits are true.

Network 3:

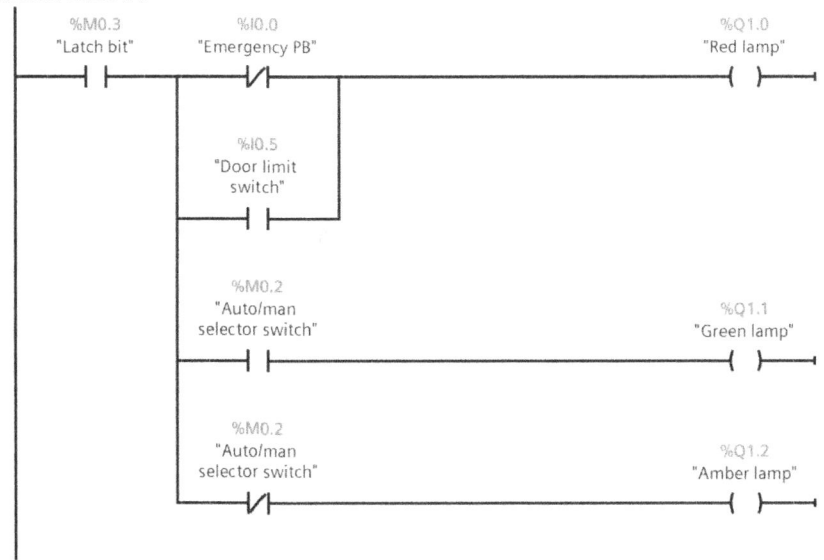

```
     %M0.3            %I0.0                                                                %Q1.0
   "Latch bit"     "Emergency PB"                                                        "Red lamp"
────┤ ├──────┬───────┤/├─────────┬──────────────────────────────────────────────────────────( )────
             │                   │
             │      %I0.5        │
             │   "Door limit"    │
             │     switch"       │
             │    ──┤ ├──────────┘
             │
             │      %M0.2                                                                  %Q1.1
             │    "Auto/man                                                             "Green lamp"
             ├──────┤ ├─────────────────────────────────────────────────────────────────────( )────
             │
             │      %M0.2                                                                  %Q1.2
             │    "Auto/man                                                             "Amber lamp"
             └──────┤/├─────────────────────────────────────────────────────────────────────( )────
```

1. A red lamp will glow only if either emergency bit or door open bit is available.

2. A green lamp will glow when machine is in auto mode.

3. An amber lamp will glow when machine is in man mode.

Network 4:

A door 1 gets opened only when part is unavailable in the cabin during auto mode. Also, when a door 1 open tab is pressed from HMI in man mode.

A door 2 gets opened after completion of hammering cycle in auto mode. And when door 2 open tab is pressed from HMI.

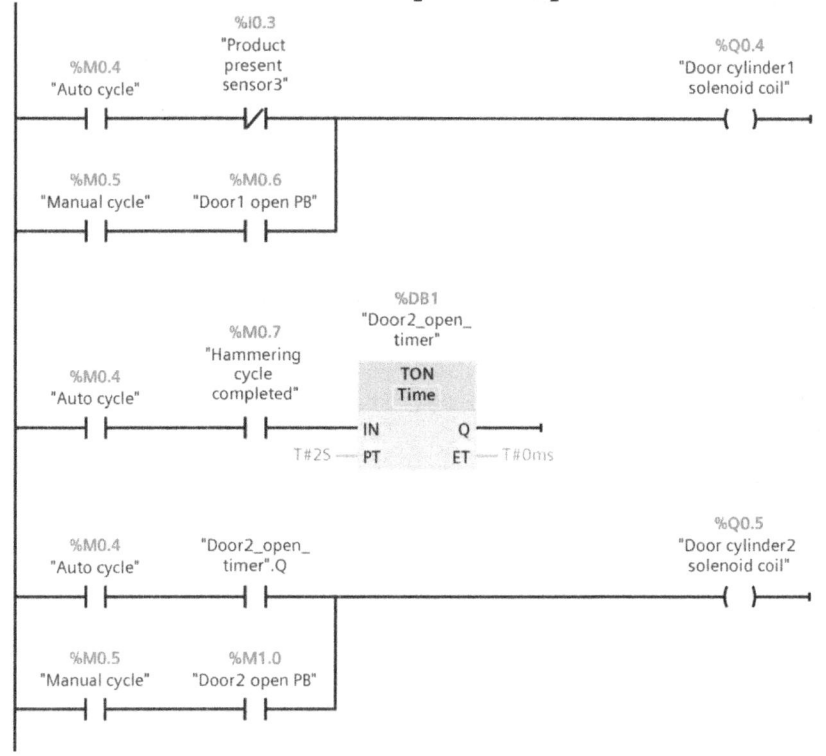

```
        %M0.4         %I0.3                                                    %Q0.4
      "Auto cycle"   "Product                                               "Door cylinder1
                     present                                                 solenoid coil"
                     sensor3"
       ─┤ ├──────────┤/├──────────────────────────────────────────────────────( )──

        %M0.5         %M0.6
      "Manual cycle" "Door1 open PB"
       ─┤ ├──────────┤ ├──

                                          %DB1
                                       "Door2_open_
                          %M0.7          timer"
                        "Hammering      ┌─────────┐
        %M0.4             cycle         │   TON   │
      "Auto cycle"      completed"      │  Time   │
       ─┤ ├──────────────┤ ├────────────┤IN      Q├──────
                                  T#2S──┤PT     ET├── T#0ms
                                        └─────────┘

                                                                               %Q0.5
        %M0.4         "Door2_open_                                          "Door cylinder2
      "Auto cycle"    timer".Q                                               solenoid coil"
       ─┤ ├──────────┤ ├──────────────────────────────────────────────────────( )──

        %M0.5         %M1.0
      "Manual cycle" "Door2 open PB"
       ─┤ ├──────────┤ ├──
```

Network 5:

Loading conveyor will start when a door is open in auto mode.

LEARN EVERYTHING ABOUT PLC PROGRAMMING

[Ladder logic diagram with three rungs:

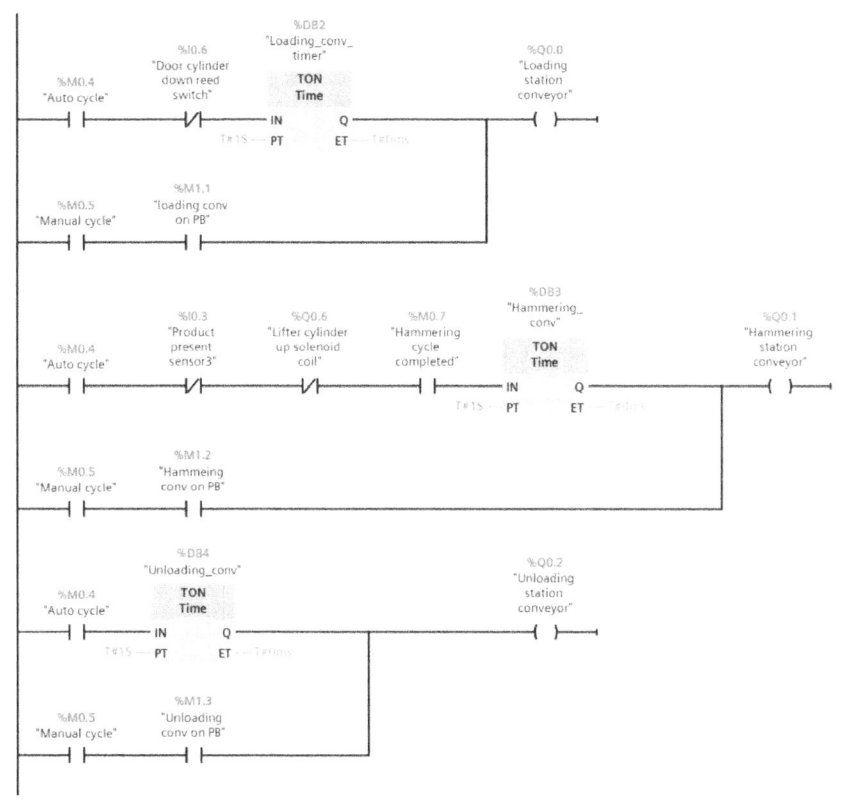

Rung 1: %M0.4 "Auto cycle" —| |— %I0.6 "Door cylinder down reed switch" —|/|— %DB2 "Loading_conv_timer" TON Time (IN, Q, PT=T#1S, ET) —( )— %Q0.0 "Loading station conveyor"
Parallel branch: %M0.5 "Manual cycle" —| |— %M1.1 "loading conv on PB" —| |—

Rung 2: %M0.4 "Auto cycle" —| |— %I0.3 "Product present sensor3" —|/|— %Q0.6 "Lifter cylinder up solenoid coil" —|/|— %M0.7 "Hammering cycle completed" —| |— %DB3 "Hammering_conv" TON Time (IN, Q, PT=T#1S, ET) —( )— %Q0.1 "Hammering station conveyor"
Parallel branch: %M0.5 "Manual cycle" —| |— %M1.2 "Hammeing conv on PB" —| |—

Rung 3: %M0.4 "Auto cycle" —| |— %DB4 "Unloading_conv" TON Time (IN, Q, PT=T#1S, ET) —( )— %Q0.2 "Unloading station conveyor"
Parallel branch: %M0.5 "Manual cycle" —| |— %M1.3 "Unloading conv on PB" —| |—]

Network 6:

The hammering operation starts when a product is inside the cabin, with doors closed and lifter is up.

The hammering time can be adjusted from HMI. A value entered from the HMI gets stored in 'hammering_operation.PT' and a timer works as per that.

215

AVINASH MALEKAR

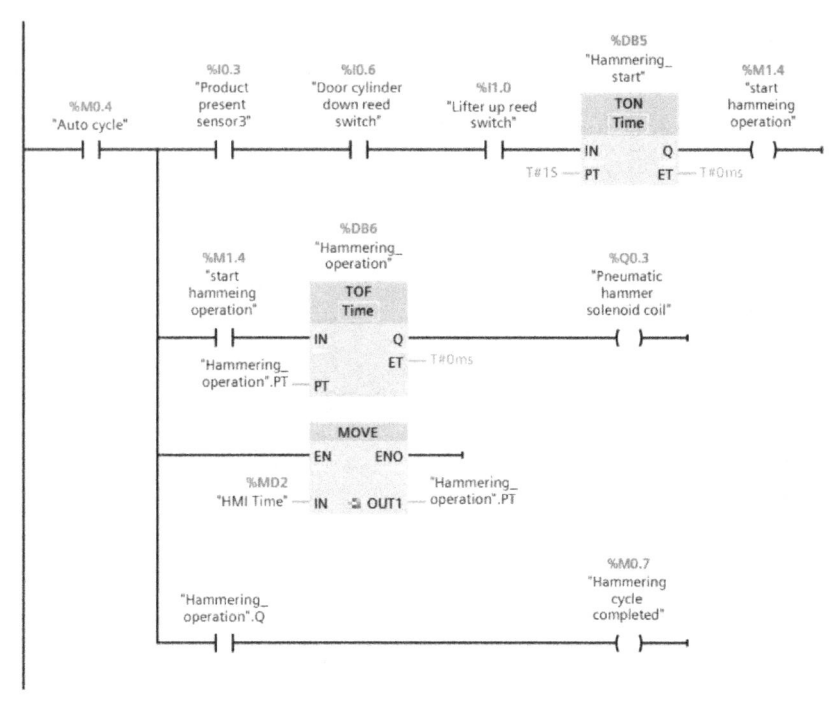

[Ladder logic diagram: Network continuation showing %M0.4 "Auto cycle", %I0.3 "Product present sensor3", %I0.6 "Door cylinder down reed switch", %I1.0 "Lifter up reed switch", %DB5 "Hammering_start" TON Time (PT T#1S, ET T#0ms), %M1.4 "start hammeing operation"; branch with %M1.4 "start hammeing operation", %DB6 "Hammering_operation" TOF Time (IN, PT "Hammering_operation".PT, ET T#0ms), %Q0.3 "Pneumatic hammer solenoid coil"; MOVE block with IN %MD2 "HMI Time", OUT1 "Hammering_operation".PT; "Hammering_operation".Q → %M0.7 "Hammering cycle completed"]

Network 7:

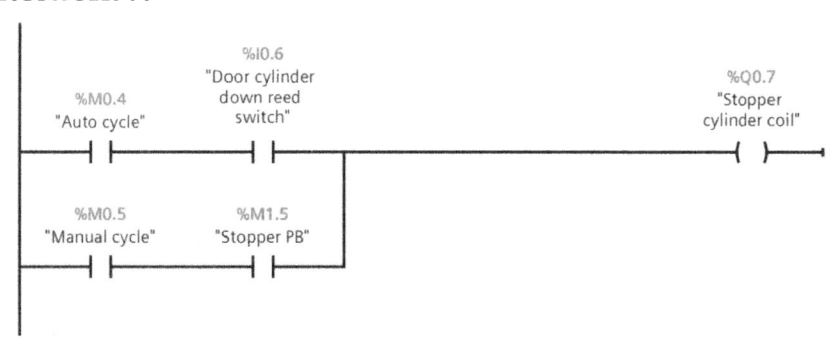

[Ladder logic: %M0.4 "Auto cycle" AND %I0.6 "Door cylinder down reed switch" → %Q0.7 "Stopper cylinder coil"; parallel branch %M0.5 "Manual cycle" AND %M1.5 "Stopper PB"]

Network 8:

Once the hammering operation gets finished, the lifter goes down after 2 sec.

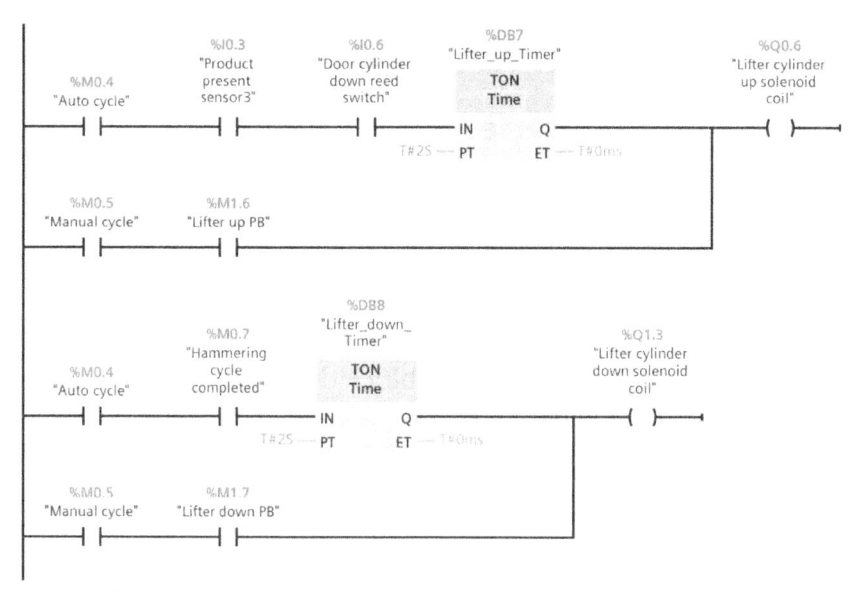

Network 9:

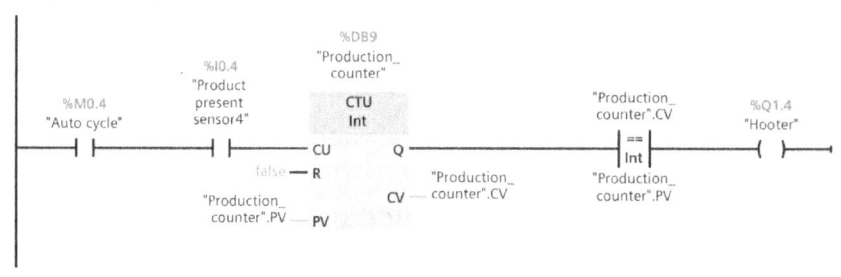

A product count sensor is provided at the exit of the hammering cabin which counts the number of products coming out of the cabin. The counter CTU counts the bits received from the sensor and stores the value in 'the production counter.CV'. An equal comparator compares the actual value with the value entered from HMI. If both values are equal, then the hooter gets on immediately.

4. MITSUBISHI PLC

A. Mitsubishi PLC series

1. MELSEC iQ-R Series

This is a high-end PLC series designed for complex automation tasks. It has a fast processing speed, a large number of input/output (I/O) channels, and advanced security features. It also supports the use of multiple programming languages, including ladder logic, function block diagram, and structured text.

2. MELSEC iQ-F Series

This is a compact and flexible PLC series designed for small to medium-sized automation tasks. It has a wide range of communication options and supports the use of standard programming languages such as ladder logic, function block diagram, and structured text.

3. MELSEC Q Series

This is a versatile PLC series designed for medium to large-scale automation tasks. It has a flexible modular design, supports a wide range of I/O modules, and offers advanced programming features such as data logging and motion control.

4. MELSEC FX Series

This is a low-cost PLC series designed for simple automation tasks. It has a small form factor, a limited number of I/O channels, and supports ladder logic programming.

B. Mitsubishi PLC programming software

These are just a few examples of the programming software available for Mitsubishi Electric PLCs. The specific software that you use will depend on the PLC series you are programming and your specific automation needs.

1. GX Works 2 and 3: This is Mitsubishi Electric's flagship programming software for its PLCs. It supports all of the

Mitsubishi Electric PLC series and provides a centralized platform for programming, simulation, and diagnostics.

Programming languages:

- Ladder logic
- Function block diagram
- Structured text.

2. GX Developer: This is another programming software for Mitsubishi Electric PLCs. It supports the MELSEC FX, Q, and L series PLCs.

Programming languages:

- Ladder logic
- Function block diagram
- Structured text.

3. GX Configurator: This is a configuration software that can be used with Mitsubishi Electric PLCs. It provides an intuitive interface for setting up the PLC, including I/O assignments and communication settings.

4. MT Developer2: This is a legacy programming software that can be used with older Mitsubishi Electric PLCs. It provides a simple interface for programming the PLC and supports multiple programming languages.

4.1 GX WORKS 3

A. Introduction

MITSUBISHI ELECTRIC MELSOFT

Programmable Controllers Engineering Software
GX Works3
Version 1

©2014 MITSUBISHI ELECTRIC CORPORATION ALL RIGHTS RESERVED

GX Works3 is the latest generation of programming and maintenance software offered by Mitsubishi Electric specifically designed for the MELSEC iQ-R and MELSEC iQ-F Series control system.

It includes many new features and technologies to ensure a trouble-free engineering environment solution.

B. Getting started with GX Works 3

1. Select type of PLC

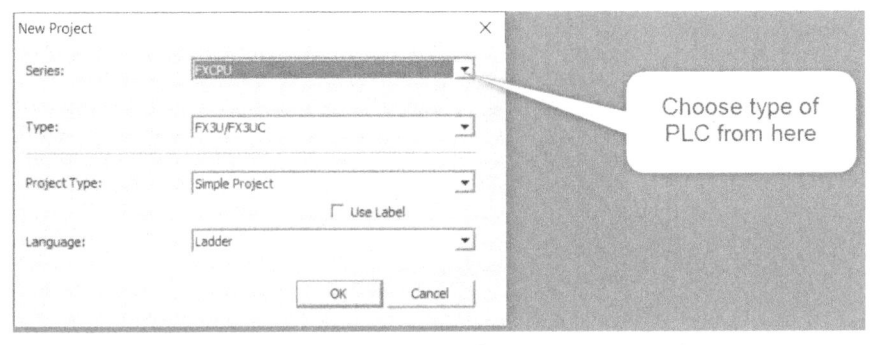

New Project dialog: Series: FXCPU, Type: FX3U/FX3UC, Project Type: Simple Project, Use Label checkbox, Language: Ladder — Choose type of PLC from here

Select the type of project. i.e. simple or structured.

Simple project- If you select simple project, then you can write a program in ladder and SFC.

Structured project- If you select a structured project, then you

can write a program in ST and FBD.
After selecting all the options, the window gets opened according to your type of language.

2. Adding a new module

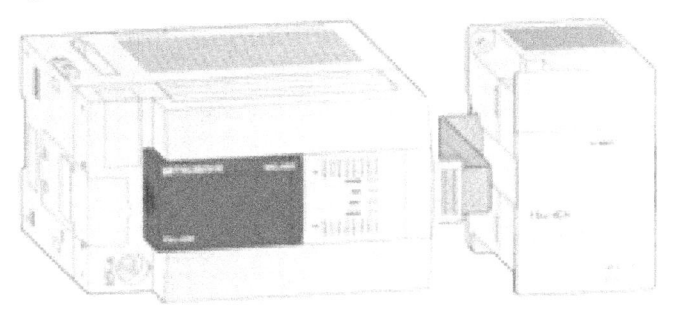

Sometimes you come with the requirements when you need to add expansion module to expand IO of your PLC or a communication module to setup different communication protocol.

If you add an Input expansion module (FX5-8EX/ES) with FX5U 32M PLC then the input increases from X20 to X27.

Parameters > FX5UCPU > module information > Right click > Add new module.

⊞ 📄 VFD
⊞ 📄 DATA_80SSC
⊞ 📄 DATA_40SSC
⊞ 📄 FAULT
⊞ 📄 AUTO_Mold
⊞ 📄 OUTPUT
⊟ 📄 PRODUCTION_REPORT
   📄 Local Label
   📄 ProgramBody
📄 Fixed Scan
📄 Event
📄 Standby
📄 No Execution Type
📄 Unregistered Program
⊞ 📄 FB/FUN
⊞ 📄 Label
⊞ 📄 Device
⊟ 📄 Parameter
   📄 System Parameter
⊞ 📄 FX5UCPU
⊟ 📄 **Module Information**
   📄 1:FX5-8EX/ES (X50)
   📄 2:FX5-8EX/ES (X60)
   📄 3:FX5-16EYT/ESS (Y50)
   📄 4:FX5-16EYT/ESS (Y70)
⊞ 📄 5[U1]:FX5-80SSC-S
⊞ 📄 6[U2]:FX5-40SSC-S
   📄 ADP1:FX5-4DA-ADP
📄 Remote Password

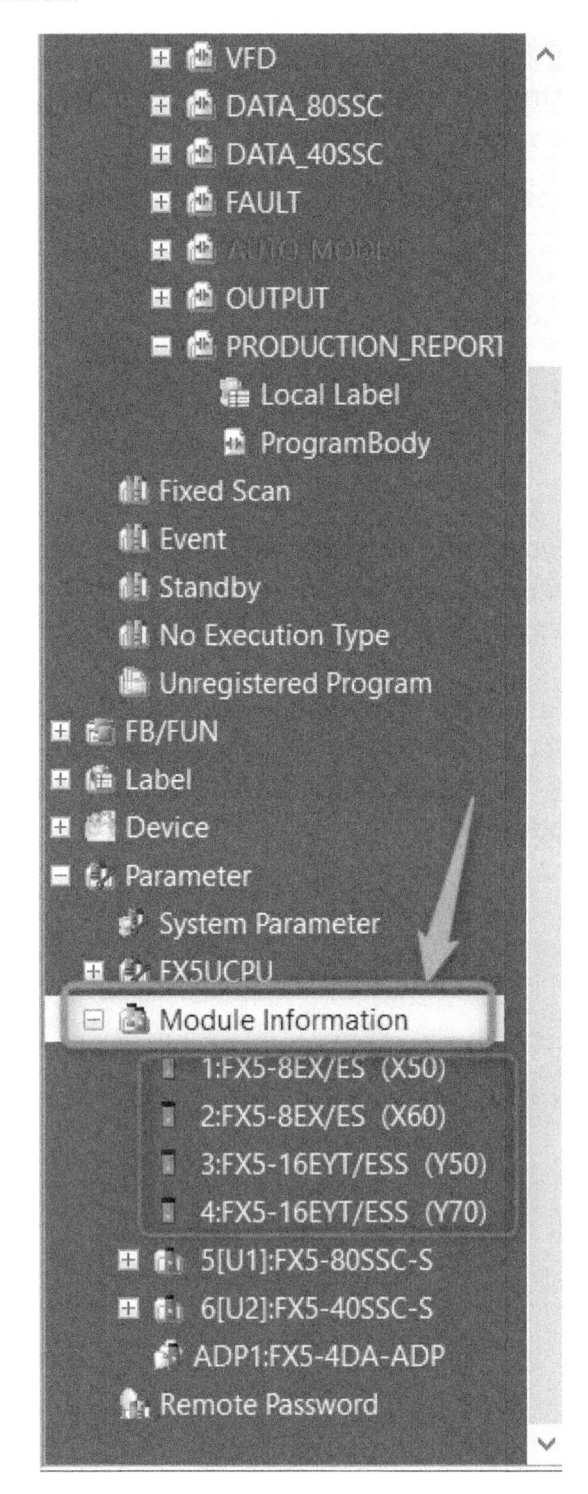

Select the module you want to add with PLC, as you can see the list has all the modules. After selecting a module, it will appear in the tree.

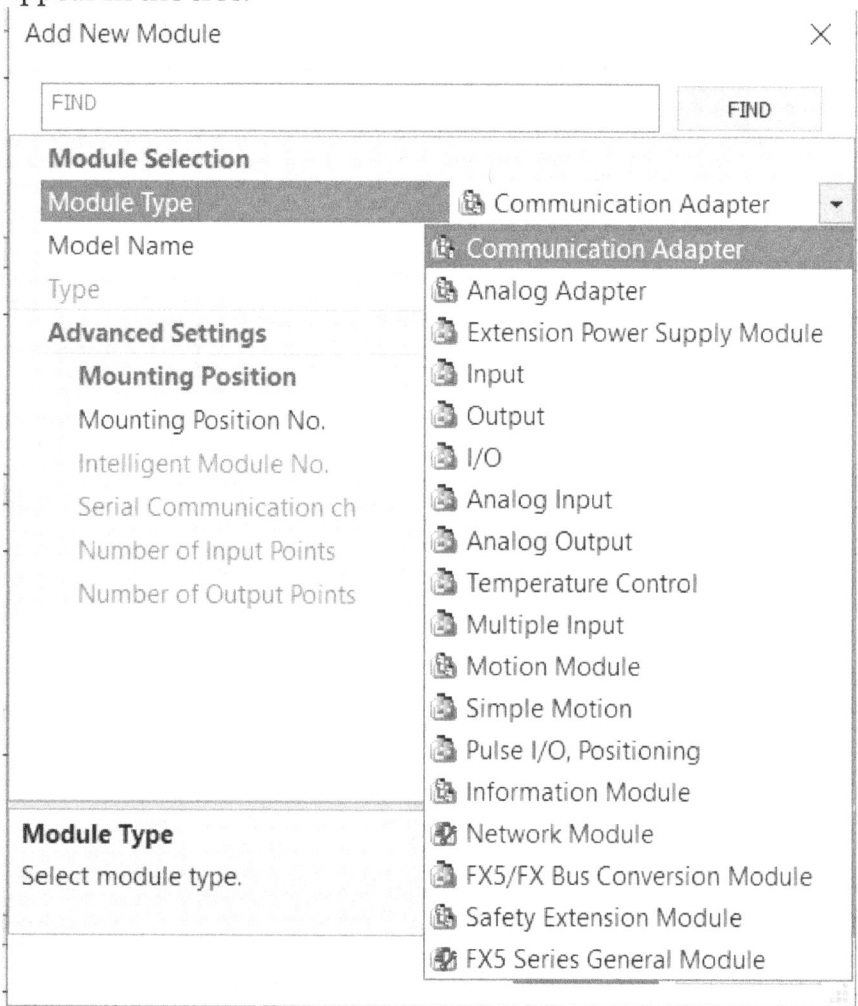

Add New Module		✕
FIND		FIND
Module Selection		
Module Type	📚 Communication Adapter	▼
Model Name	📚 Communication Adapter	
Type	📚 Analog Adapter	
Advanced Settings	📚 Extension Power Supply Module	
Mounting Position	📚 Input	
Mounting Position No.	📚 Output	
Intelligent Module No.	📚 I/O	
Serial Communication ch	📚 Analog Input	
Number of Input Points	📚 Analog Output	
Number of Output Points	📚 Temperature Control	
	📚 Multiple Input	
	📚 Motion Module	
	📚 Simple Motion	
	📚 Pulse I/O, Positioning	
	📚 Information Module	
Module Type	📚 Network Module	
Select module type.	📚 FX5/FX Bus Conversion Module	
	📚 Safety Extension Module	
	📚 FX5 Series General Module	

3. Program writing procedure
1. Overview screen

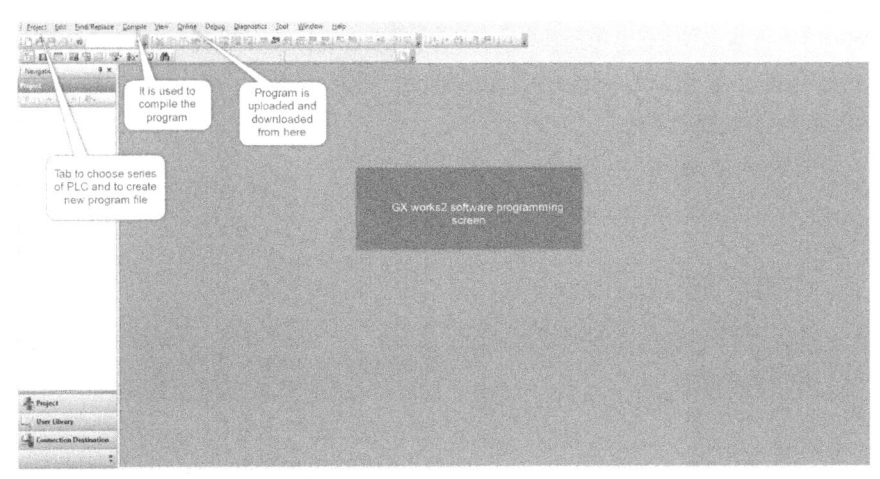

2. PLC programming languages

Simple project- If you select simple project, then you can write a program in ladder and SFC.

Structured project- If you select a structured project, then you can write a program in ST and FBD.

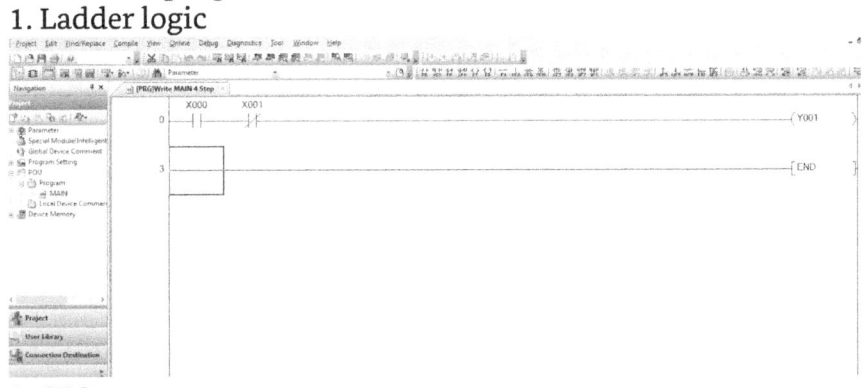

1. Ladder logic

2. SFC

LEARN EVERYTHING ABOUT PLC PROGRAMMING

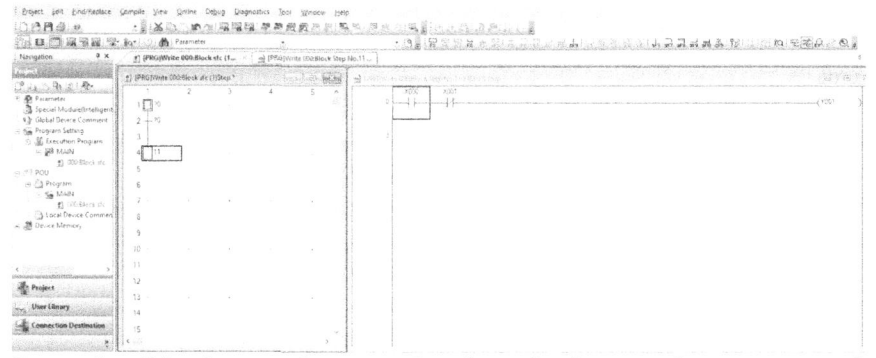

Function block- An element that has the internal memory and output an operation result according to values in the memory and the input value.

Function- An element that has no internal memory and output always the same operation result for same input value.

3. ST

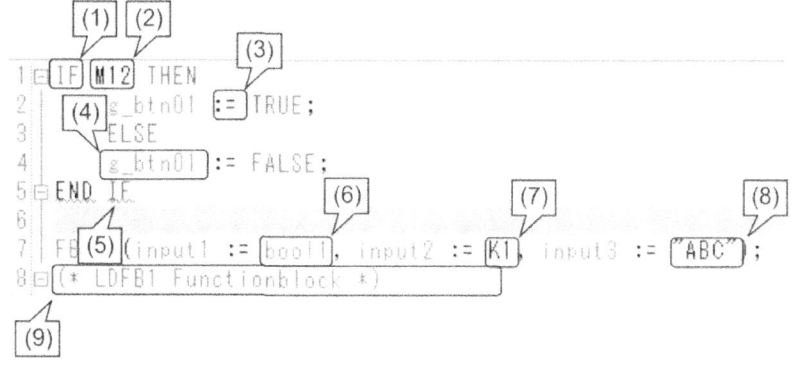

3. Write a PLC program

Ctrl + navigation arrow to create a rung like shown above.
Write a device command in the description box.

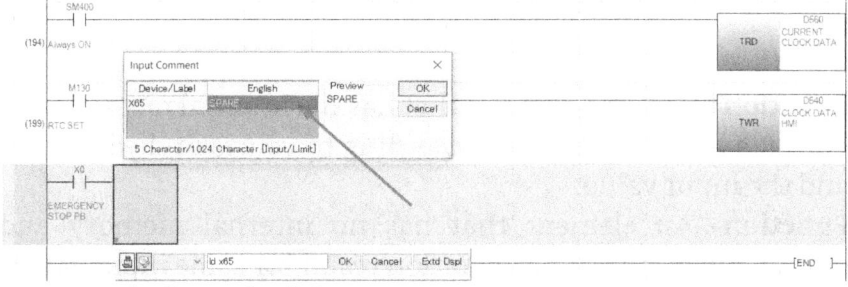

4. Screen before rebuild

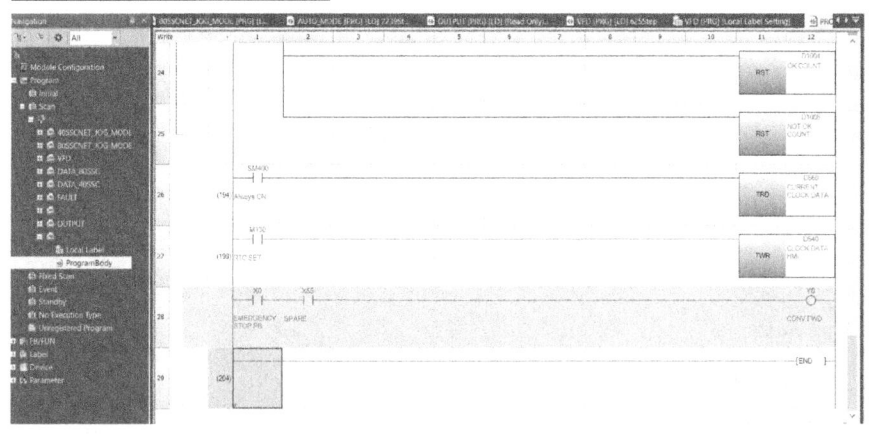

5. Screen after rebuild

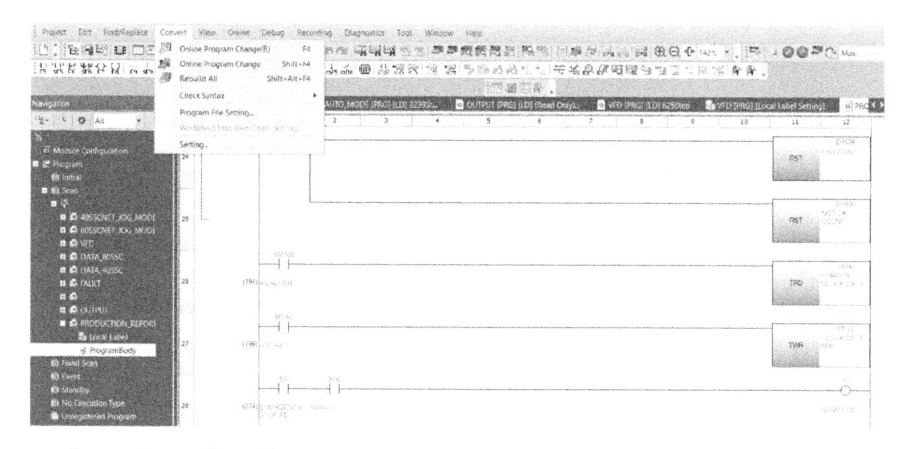

Before downloading a program into PLC, you have to make sure it is rebuilt, otherwise you cannot download it.

3. Communication with PLC and PC

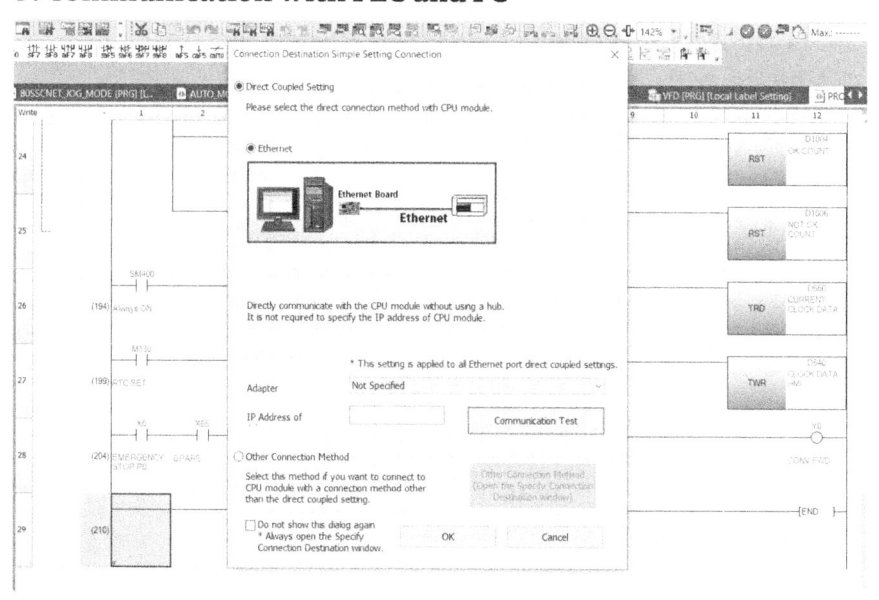

You do not need extra software to establish a communication between PLC and your computer. You can establish communication with two cables:
1. USB cable
2. Ethernet cable

Ethernet connection

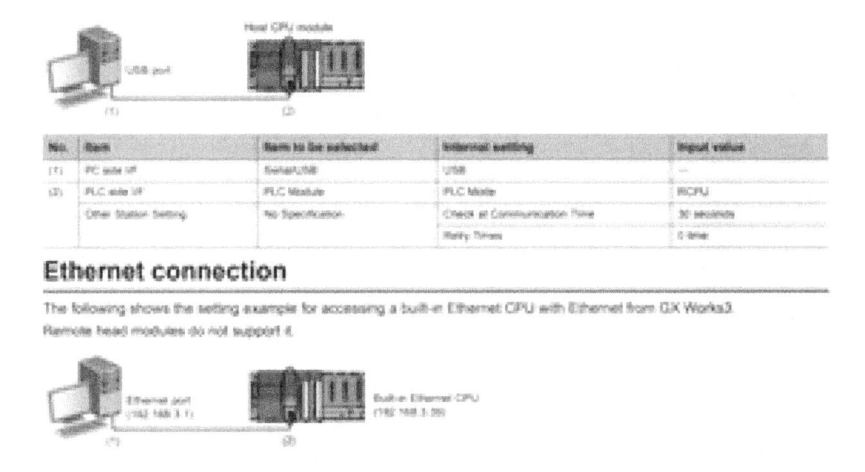

The following shows the setting examples for accessing a built-in Ethernet CPU with Ethernet from GX Works3.
Remote head modules do not support it.

Make sure the IP addresses of your PC and PLC is in series to establish a connection. You can click on connection test to check their connection.

4. Upload and download PLC program

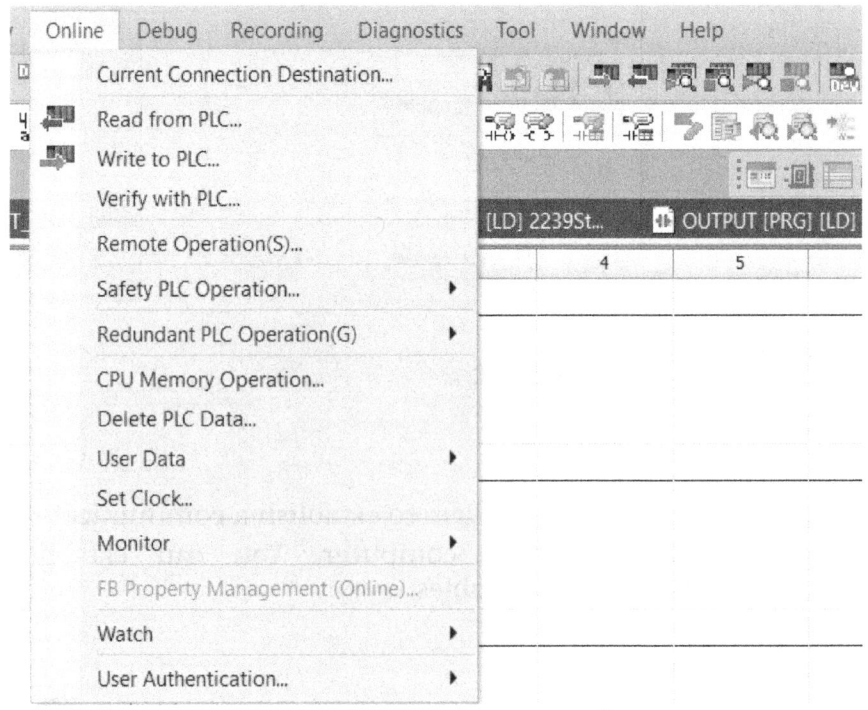

Read from PLC- To upload a program from PLC to your computer.

Write to PLC- To download a program from your computer to PLC.

C. GX Works 3 ladder instructions
1. Digital signals

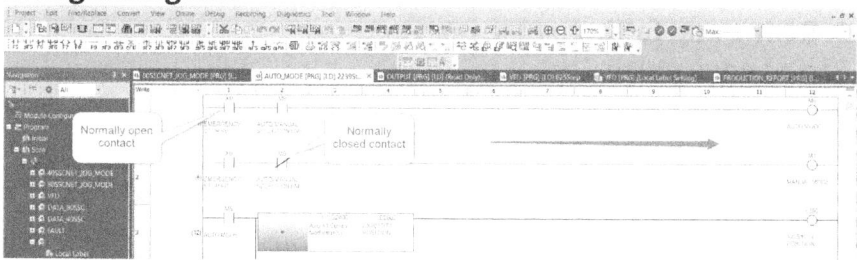

Input (X): Inputs are designed to give commands and data from external devices. e.g. pushbuttons, select switches, etc., to the PLC.

Output (Y): Outputs are used to provide the control results of a program from the PLC to external devices, e.g. solenoids, signal lamps.

Internal relays (M): Internal relays are auxiliary relays used in a CPU module and not latched.

Latch relays (L): Latch relays are auxiliary relays used in a CPU module and latched (backed up at power failure)

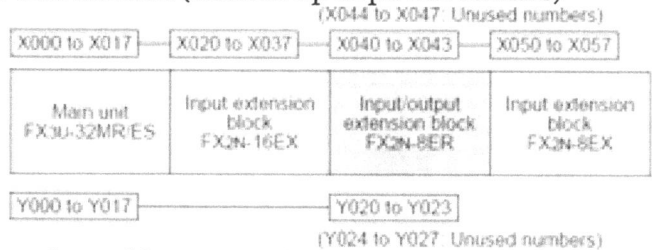

1. Addressing of inputs and outputs is in Octal (X0-X7, X10-X17)
2. Addressing for both inputs and outputs start at 0 (X0 and Y0)
3. Addressing is consecutive.

If input/output powered extension units/blocks have been connected when the power is turned on, the main unit automatically assigns the input/output numbers (X/Y) (octal) to the units/blocks. Therefore, it is unnecessary to specify the

input/output numbers with parameters.
We have chosen FX3U PLC in which digital inputs start from X00, X01, X02, X03, X04, X05, X07, X10, X11....
And outputs start from Y00, Y01, Y02, Y03, Y04, Y05, Y06, Y07, Y10, etc.

2. Rising and falling pulse instructions
Rising edge pulse

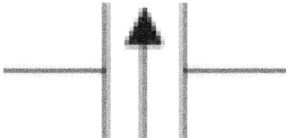

True for one scan on up transition at address

1. Operating a lamp from a push button
When a push button is pressed the first time a lamp will glow. When a push button is pressed a second time a lamp will turn off.

2. Home automation

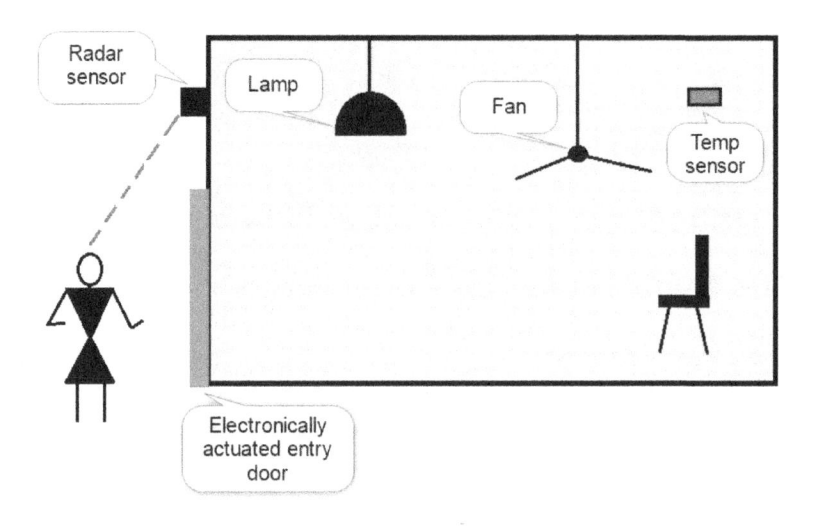

A controller is used to control the door, lamp, and a fan inside the room. A radar sensor is provided at entry of the room which detects the entry on an object or a human. When it detects anything, it sends a signal to the controller and the door starts sliding to the left side after 1 sec and gets closed after 3 sec. As soon as an object enters the room, the lamp will glow.

If the temperature inside the room is higher than 38 deg cel, a fan will start. A digital temperature sensor is used to detect temperature inside the room, it sends a signal to the controller if the temperature is greater than 65 deg cel.

Falling edge pulse

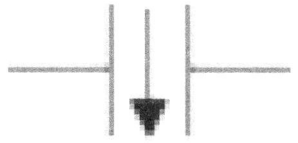

True for one scan on the down transition at address

4. Timer

As we already know a timer instruction is used to delay in operation or to control the operation.

	T1	K300
OUT	Door control timer	

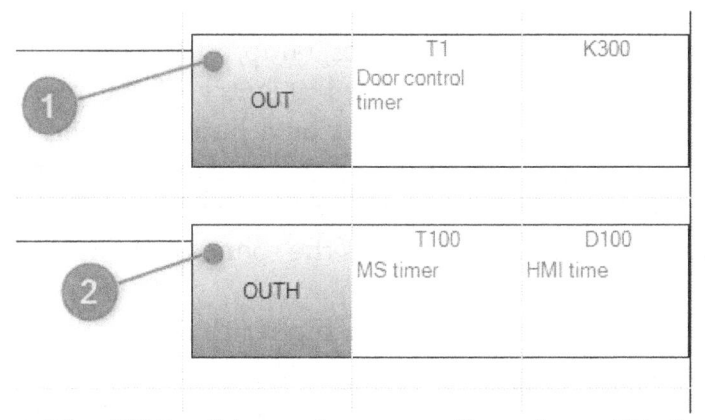

	T100	D100
OUTH	MS timer	HMI time

1. The OUT coil is used to prescribe a timer. T1 shows the type of timer base, it could be in seconds or ms. Data register K is used to store the timer value.

2. The OUTH is used for high speed timer. T100 shows th e the type of timer base. Data register D is used to stored the timer value. We assign the data register only in case of HMI. In that case we can change the timer's value from HMI.

Timer Base

The PLC timer base depends on the series of PLC. For example, if you wish to use a timer which has time base in ms, in that case you can choose a timer that lies within 0-199 for FX 3U PLC series.

Time Base	FX1S	FX1N/FX2N/FX2NC	FX3U
100ms	0-62	0-199	0-199
10ms	32-62	200-245	200-245
1ms (Retentive)	-	246-249	246-249
100ms (Retentive)	-	250-255	250-255
1ms	63	-	256-511

To operate this timer, you have to make sure a connected input is always true in case of NO switch and should be false in case of NC switch. Mitsubishi does provide an extra time instruction for off delay timer. A same timer instruction can be used as a on delay and off delay timer. Let's understand the use of a timer as on delay and off delay applications.

1. Conveyor application

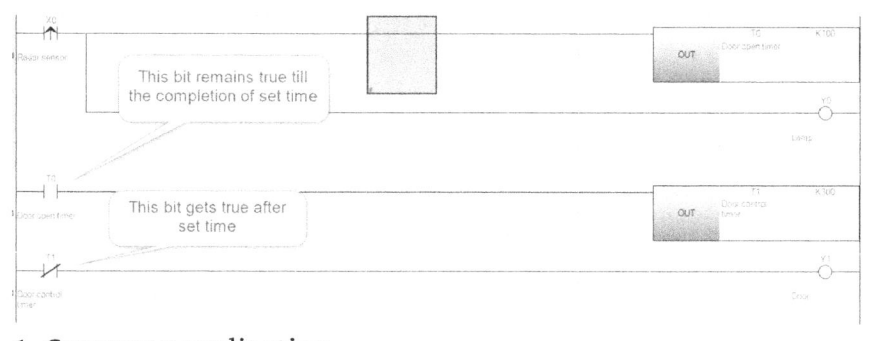

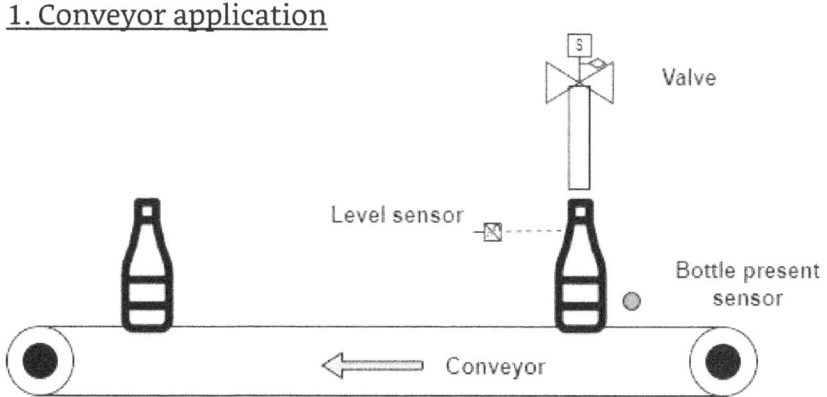

A bottle filling line is as shown above. It consists of a flat belt conveyor which carries bottles and a filling station where the bottles are filled. The conveyor is controlled by a VFDs and valve is controlled by a solenoid coil.

When the sensor detects the presence of the bottle at the filling station the solenoid valve will start filling the bottle until the level sensor is true. This cycle will continue till stop PB is pressed.

Input devices
- Start PB (X00)
- Stop PB (X01)
- Bottle present sensor (X02)
- Level sensor (X03)

Output devices
- Conveyor motor (Y0)
- Hopper motor (Y01)

Programming using works3

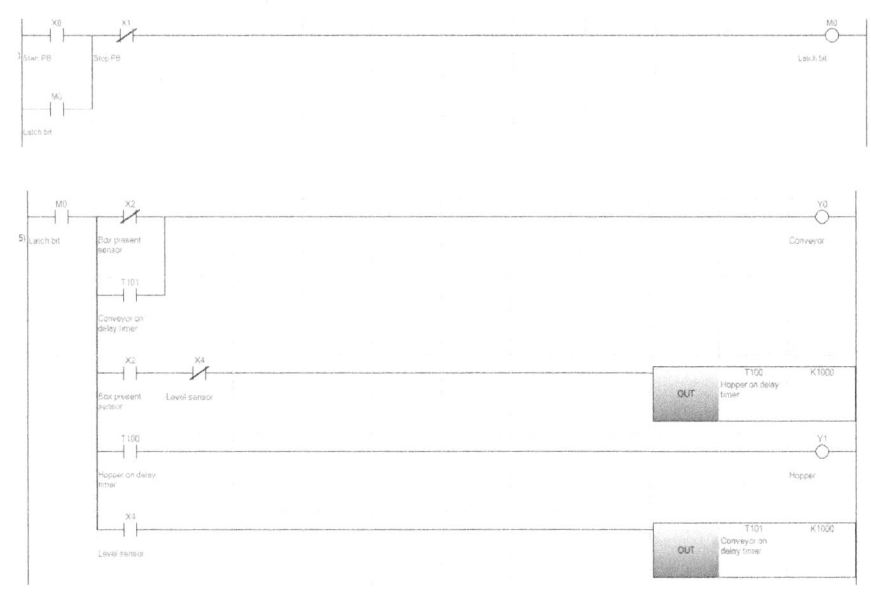

2. Flash light

When a button is pressed, a lamp gets on for 1 sec and gets off for 0.5 sec. The lamp will continue flashing for 20 sec.
A lamp will not flash after pressing PB 5 times, it will reset by reset button only.

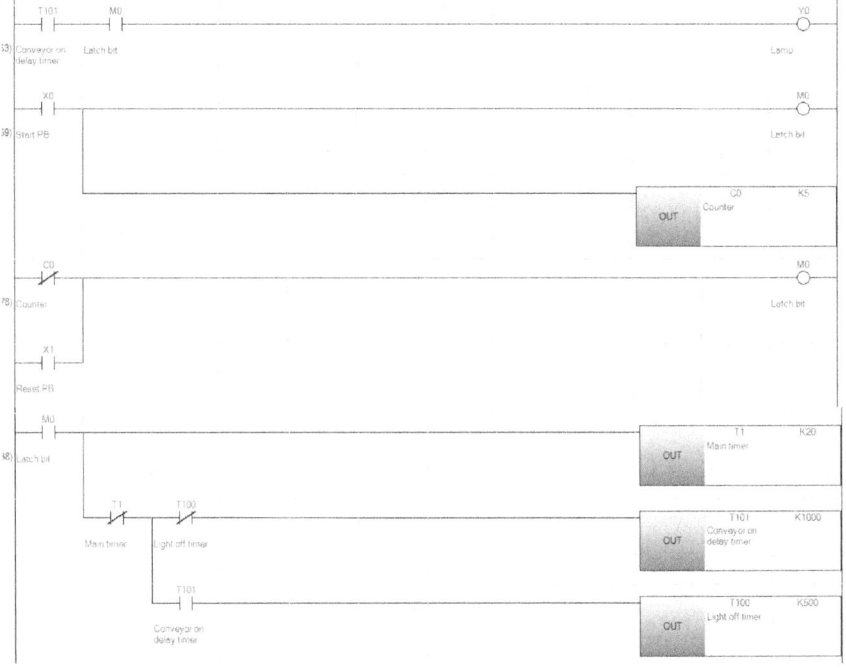

5. Counter

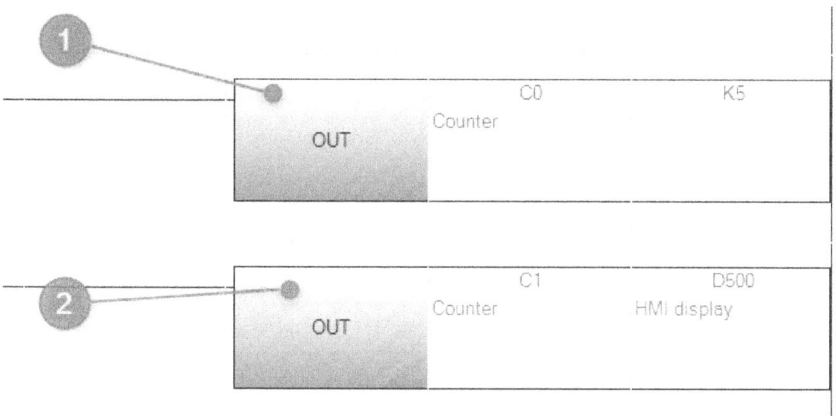

①
```
                        C0              K5
                   Counter
         OUT
```

②
```
                        C1              D500
                   Counter          HMI display
         OUT
```

It is also an output instruction. This counter is retentive and a reset instruction is used to reset it. This counter is available in 16/32-bit.

1. C0 is a counter which counts the value which is stored in data register k.
2. C1 is a counter which counts the value which is stored in data register D. I have assigned this data register to the display tab on HMI which shows the counting value on HMI directly.

```
  X0                                              C0      K5
──┤├─────────────────────────────────────────OUT  Counter
  Start PB

  X1                                                      C0
──┤├─────────────────────────────────────────RST  Counter
  Reset PB
```

Understand the following:
Preset: Constants or Register.
Accumulator: Current status of the counter, use the address of Counter Device.
Contact or Done bit: Can be accessed using the address of Counter Device.

Production report

When the start PB is pressed after every 4 seconds a red marker is manufactured, and after every 7 seconds a black marker is manufactured.

When total markers are manufactured in 7 and 9 quantities then the yellow lamp and red lamp will glow respectively.

Programming using works3

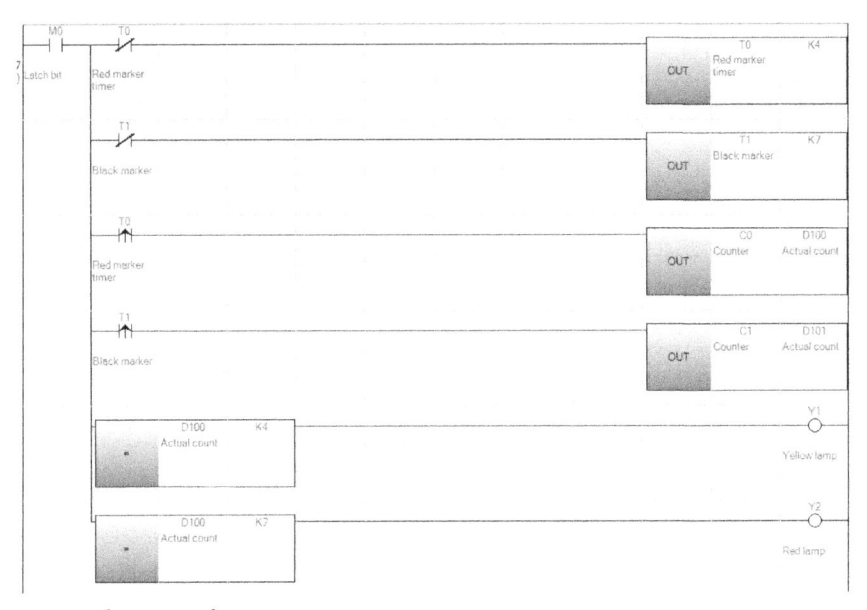

[Ladder diagram with inputs X0 (Start PB), X1 (Stop PB), M0 (Latch bit)]

[Ladder diagram with M0 (Latch bit), T0, T1 controlling Red marker timer (K4), Black marker (K7), Counter C0 (D100), Counter C1 (D101), and outputs Y1 (Yellow lamp), Y2 (Red lamp)]

6. Math Functions

[Ladder diagram showing M0 (Latch bit) with addition and subtraction operations on D100 (Actual count) and D101 (Allowance)]

The math fuctions are used to perform different calculation during programming. It covers almost all types of math fuctions, but generally we use:
1. Addition
2. Subtraction
3. Multiplication
4. Division

8. Compare instructions

236

Instruction compares two data and results in a true or false condition. It can be used with other sequence instructions. All the instructions are input instructions except CMP which is an output instruction.

Inline comparisons

The data of C0 is compared to the data of K0, and the result gets displayed accordingly.

–K1 is less than C0 - bit device Y01 is ON
–K1 is equal to C0 - bit device Y00 is ON

D. Troubleshooting

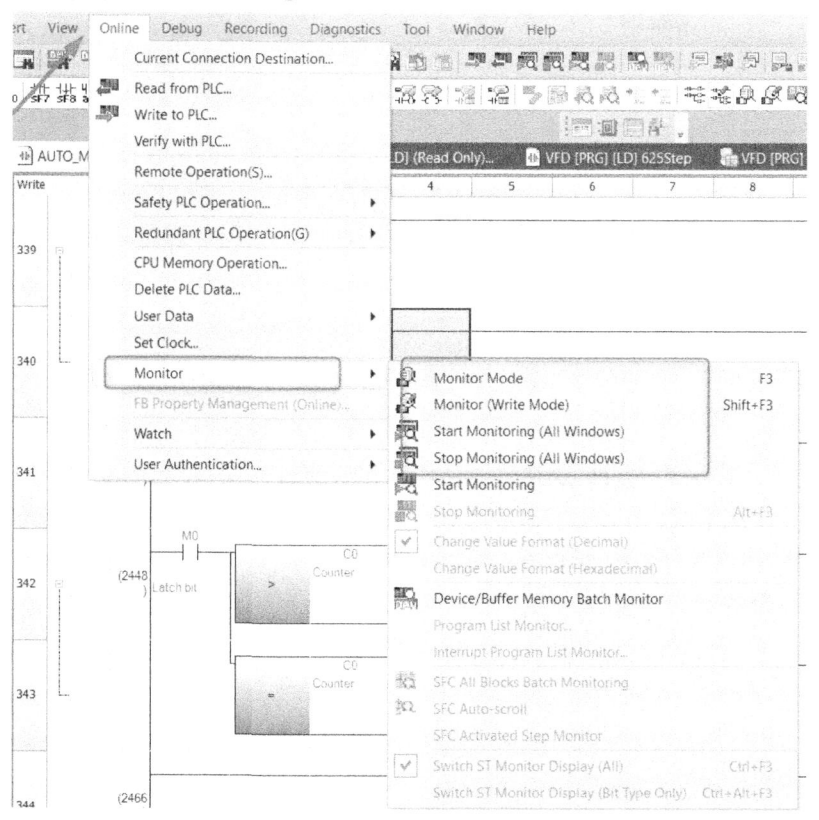

Monitoring mode- In this mode you can just check the status of IO devices, but the program cannot be modified.

Monitoring & write mode- in this mode you can not only observe the IO devices status but also make some changes in program without disturbing machine's operation.

Online program change- After making changes in program, click on convert option and then select online program change. Let's consider your machine gets broken down during operation and you observe its pusher cylinder is not working. You can troubleshoot by online monitoring of a program. Online monitoring takes less time as compare to physical tracing of wires.

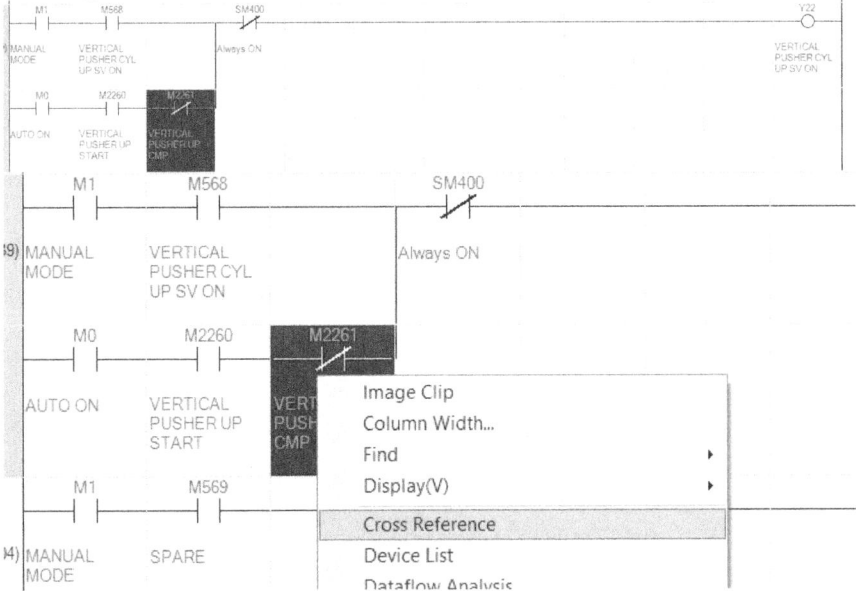

This type of chart will appear in front of you. It shows the use of M2261 bit in different rungs.

To make this rung true, M2261 should be true. Right click on in and select cross reference.

LEARN EVERYTHING ABOUT PLC PROGRAMMING

Device/Label	Device	Ladder Sy...	Position	Program File Name	Data Name	English	Access from Extern...
M0	M0		485 Serial Port		FX5UCPU	AUTO ON	
M2261	M2261	-|/|-	1031 Step	MAIN	AUTO_MODE	VERTICAL P...	
M2261	M2261	-|/|-	1053 Step	MAIN	AUTO_MODE	VERTICAL P...	
M2261	M2261	-( )-	1151 Step	MAIN	AUTO_MODE	VERTICAL P...	
M2261	M2261	-| |-	1148 Step	MAIN	AUTO_MODE	VERTICAL P...	
M2261	M2261	-| |-	1159 Step	MAIN	AUTO_MODE	VERTICAL P...	
M2261	M2261	-|/|-	297 Step	MAIN	OUTPUT	VERTICAL P...	

For troubleshooting you need to select a coil of M2261, after selecting a coil from above chart a new rung will appear, you can check which signal is pending and take an action. Also, you can use **shift + enter** to make a particular coil forcefully true.

E. GOT 2000 HMI

In this section we are going to cover the PLC programming related to HMI only. I have already created HMI program using GT Designer 3 as shown above.

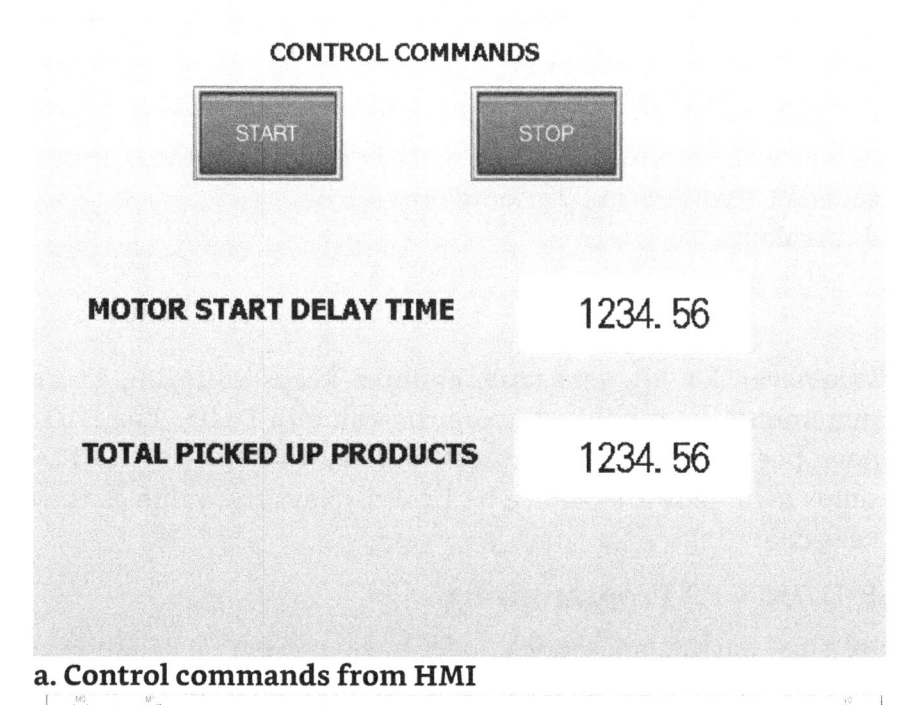

CONTROL COMMANDS

START   STOP

MOTOR START DELAY TIME   1234. 56

TOTAL PICKED UP PRODUCTS   1234. 56

a. Control commands from HMI

A machine can be easily control through HMI by creating

239

control tabs. As you can see above, two tabs have been created and memory bits M0 and M1 have been assigned as per the program.

b. Numeric entry from HMI

HMI provides with a facility to change the values of parameters such as Frequency, time, etc. To accept the value from outside we need to use numeric input tab.

As the above rung shows, the time of motor start delay can be changed by entering value from HMI.

c. Alarm and notification

As per the above rung, if X1 is present the timer starts counting 2 seconds. If the product is not picked up within 2 seconds, then Alarm Y2 should on.

d. Display

Whenever X1 bit gets true, counter keeps counting in an incremental manner and store the value in D101. The D101 have been assigned to numeric display tab on HMI, so PLC sends a value to the HMI. The HMI displays the value as they keep changing.

F. D700 VFD Programming

In this particular section, we have chosen a Mitsubishi D700 VFD which is controlled by FX5U with Modbus communication. In this section we will cover:

a. Mitsubishi FX5U and D700 VFD wiring diagram

LEARN EVERYTHING ABOUT PLC PROGRAMMING

GOT 2000

FX 5U

ETHERNET

D700 VFD

MODBUS

RDB		SDB
SDB		RDB
RDA		SDA
SDA		RDA
SG		SG

FX 5U MODBUS
TERMINALS

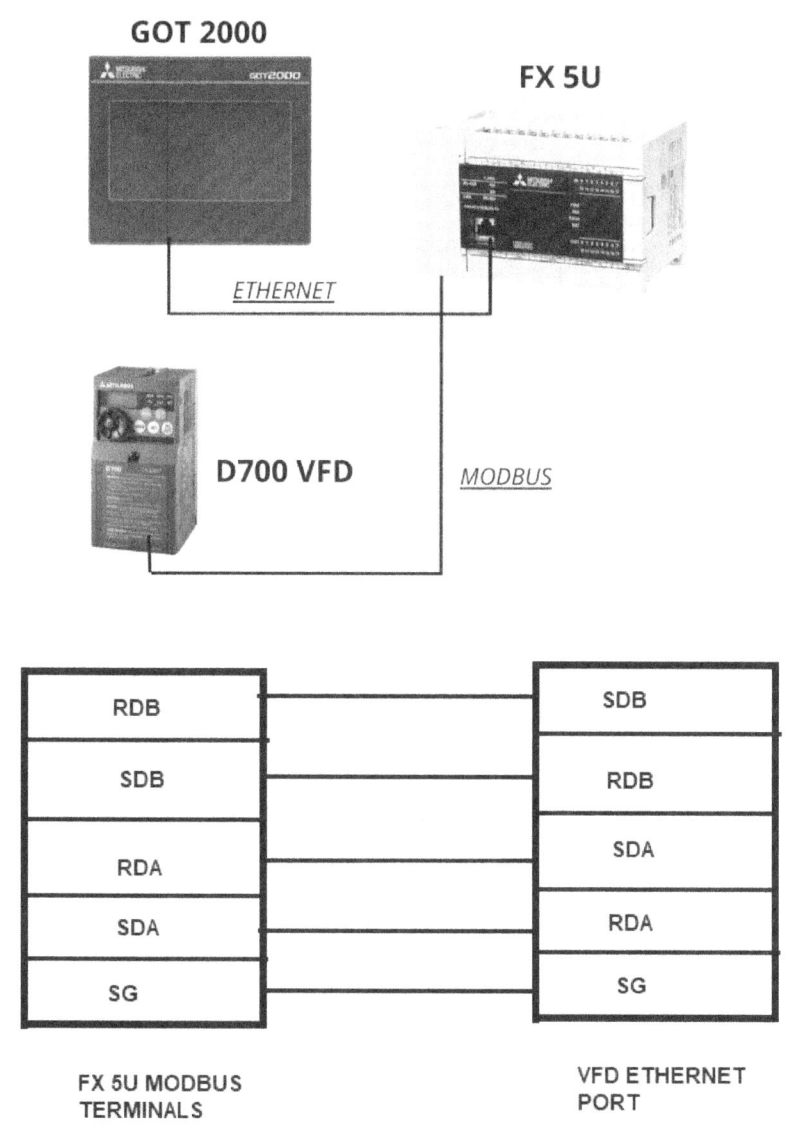

VFD ETHERNET
PORT

b. D700 VFD Modbus configuration
We need to set the communication parameters as the table below to be able to control D700 with Modbus RTU.

Parameter no	Description	Set value
P77	Parameter write selection	2
P79	Operation selection mode	2
P117	PU communication station no	1
P118	PU communication speed	192
P119	PU communication stop bit length	1
P120	PU communication parity check	2
P121	Number of PU communication retries	9999
P124	PU communication CR/LF presence/ absence selection	1
P338	Communication operation command source	1
P339	Communication speed command source	0
P340	Communication startup mode selection	1
P549	Communication EEPROM selection	1
P549	Protocol selection	0
P551	PU operation command source selection	9999

c. Mitsubishi FX5U Modbus RTU Setting

We need to configure Modbus communication parameters in PLC same as D700 VFD.

LEARN EVERYTHING ABOUT PLC PROGRAMMING

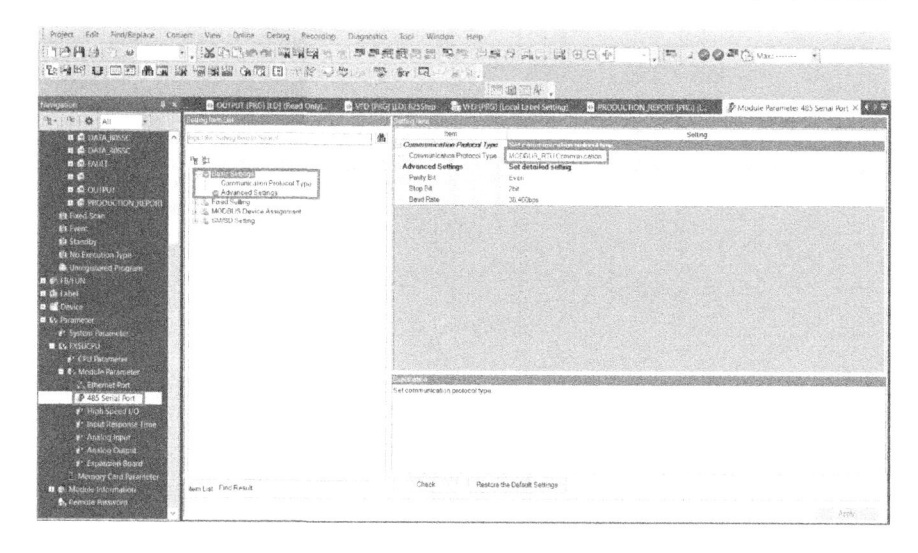

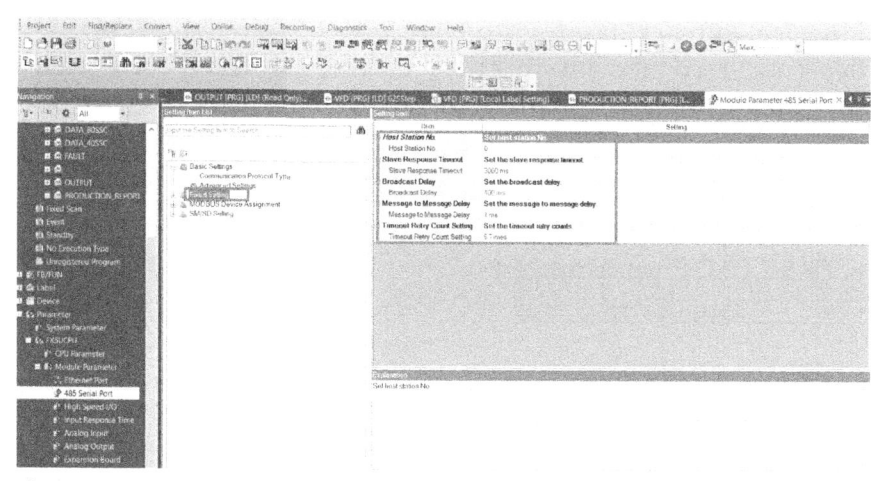

d. GOT 2000 HMI Screen

243

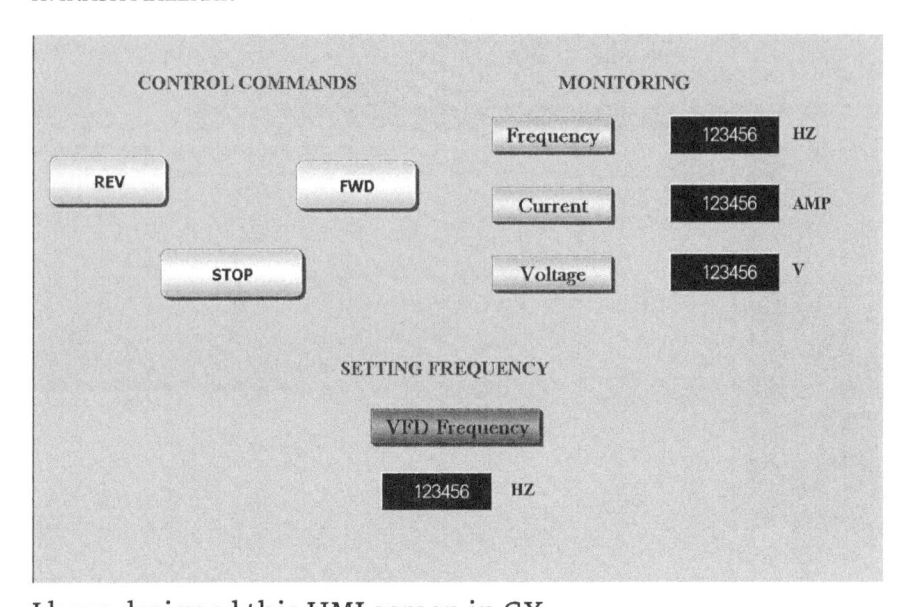

[HMI screen showing CONTROL COMMANDS (REV, FWD, STOP buttons), MONITORING (Frequency, Current, Voltage with 123456 displays in HZ, AMP, V), and SETTING FREQUENCY (VFD Frequency, 123456 HZ)]

I have designed this HMI screen in GX designer. Two tabs have been provided to change the direction of motor. You can set the frequency the VFD. Also you can monitor different parameters such as running frequency, current, and voltage over HMI.

e. FX5U PLC Programming
Read / Write Distribution

[Ladder logic diagram showing SM8002 After RUN 1 Scan ON, with MOV K0 D20 SPEED INPUT and MOV H43C D30 BASE SCREEN SWITCHING]

Reset control when PLC stop

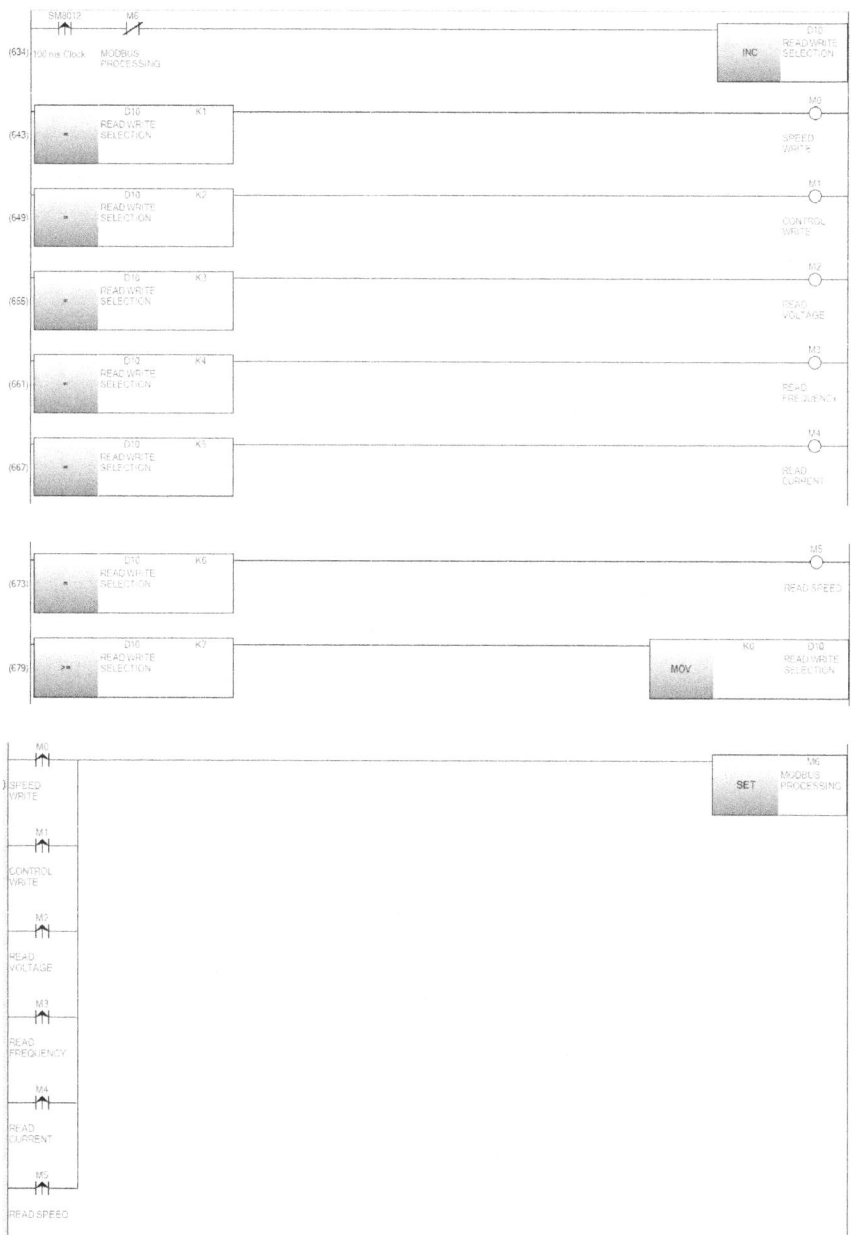

Speed Setting Command

The data register D20 stores frequency which is entered from HMI and then sends it to VFD.

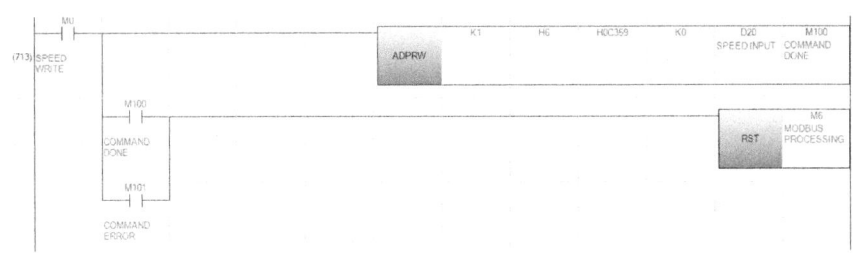

VFD Control Run

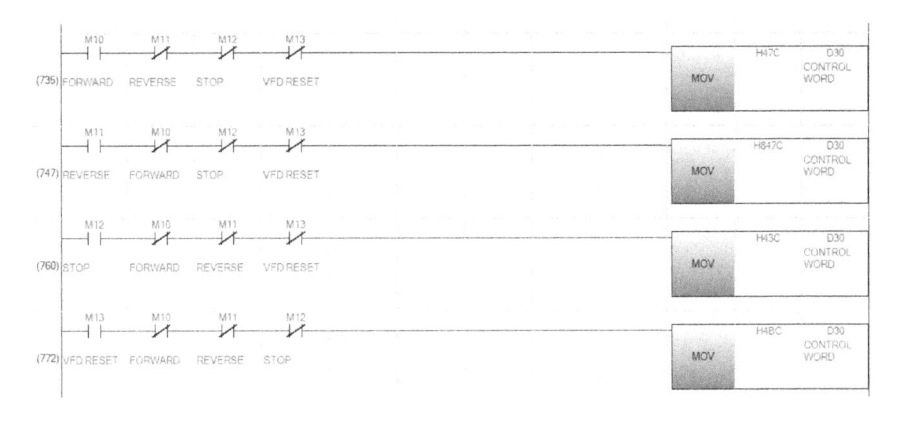

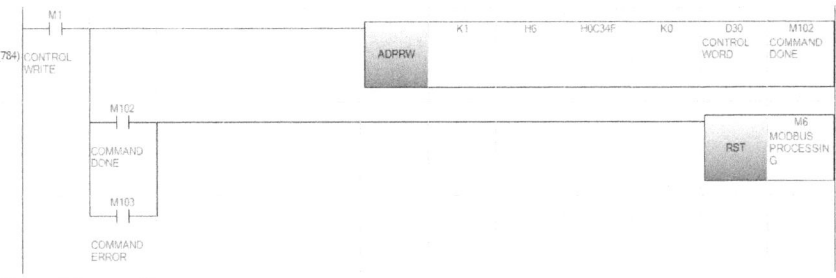

Read Data Response

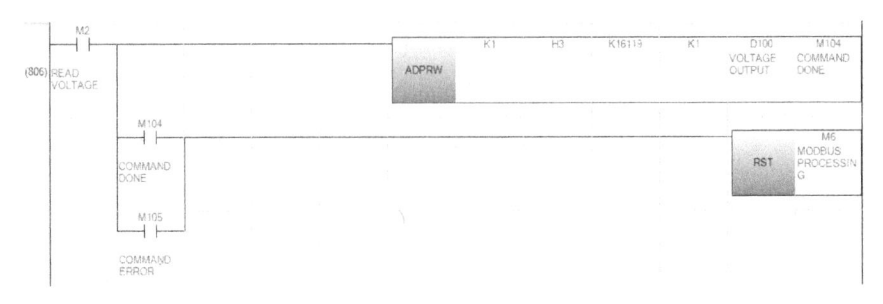

LEARN EVERYTHING ABOUT PLC PROGRAMMING

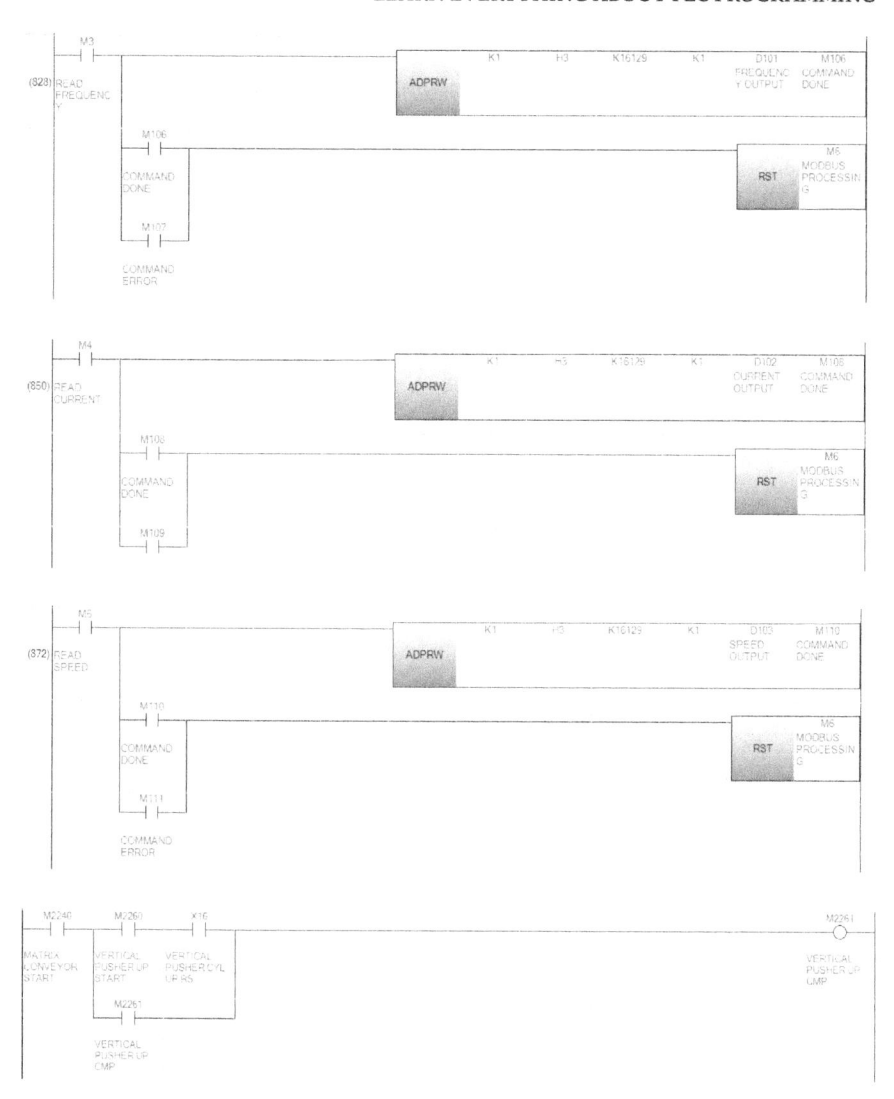

AVINASH MALEKAR

CASE STUDY 4

Assembly line for road roller.

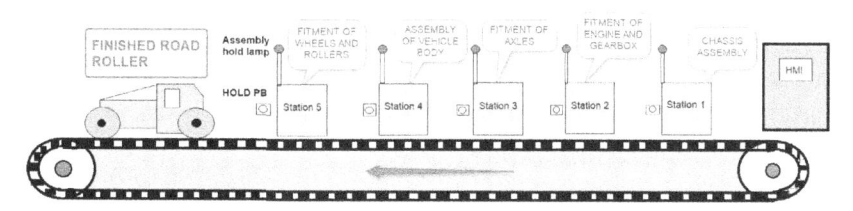

An assembly line includes fitting of engine on chassis, fitment of axles, mounting of body, and fitment of wheels and rollers. The conveyor is driven by an electric motor coupled with a gearbox along with an electric brake.

Hardware and software details
PLC - FX 3U 32M
HMI- GOT 2000
PLC & HMI communication- Ethernet
PLC Programming software- GX Works3
HMI Programming software- GT Designer3

<u>Input devices</u>
Emergency PB (X00)
Start PB (X01)
Stop PB (X02)
Station-1 Pause PB (X03)
Station-2 Pause PB (X04)
Station-3 Pause PB (X05)
Station-4 Pause PB (X06)
Station-5 Pause PB (X07)
Finished goods sensor (X10)

<u>Output devices</u>
Conveyor forward command (Y00)
Station 1 hold lamp (Y01)
Station 2 hold lamp (Y02)
Station 3 hold lamp (Y03)

Station 4 hold lamp (Y04)
Station 5 hold lamp (Y05)
Conveyor motor brake (Y06)
Conveyor stop command (Y07)
Conveyor reverse command (Y10)
HMI Commands
Auto/Man tab (M09)
Conveyor forward tab (M04)
Conveyor reverse tab (M05)

Operation

1. When a start PB is pressed, a motor brake will be released first and after 1 sec the conveyor forward command will be given to the VFD. A conveyor will move in forward direction with 2 mm/1 sec speed.

2. Assembly of components is done during forward movement of the conveyor. In case an operator requires more time, he can pause the conveyor and finish his work, after releasing the pause button the conveyor will restart. All 5 stations have been provided with a pause button. The notification should be displayed on HMI.

3. When a stop PB is pressed, the conveyor will be stopped and after 1 sec a brake will be applied. The speed of the conveyor should be settable from HMI. The production report should be generated on HMI.

GOT 2000 HMI Screen

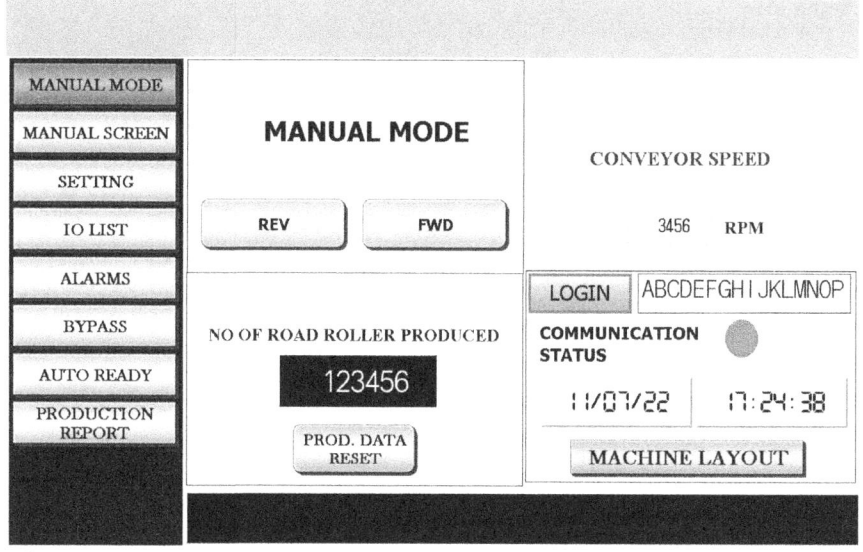

The screen has been created using GT Designer3 HMI programming software. Let's consider the necessary memory bits have been assigned to the respective tabs.

This screen can be used for-

-Control the forward and reverse direction of the conveyor

-To change the speed of the conveyor

-To check the production report

Programming using GX Works 3

1. Create a project

- Select the series of PLC
- Select PLC type
- Select program language

1. Auto mode

Rung 1- M0 bit is assigned to the HMI tab, and it gets true when it is pressed.

Rung 2- Cycle start memory bit gets latched when all pause buttons are false. Latching bit is provided parallel to the X01.

Rung 3 & 4- Once the start command is received, it will release the brake first and switch one the motor after 1 sec. A high-speed timer OUTH has been used here.

Rung 5- All the pause buttons are parallel to each other which means, the conveyor stop command gets activated when any one of the button is pressed.

Rung 6 & 7- Once the stop command bit gets latched, it will stop the motor first and then apply the brake after 1 sec.

2. Stop indication

Rung 8 to 12- The station pauses lamp glows as soon as pause button is true.

3. Manual mode

Rung 13 to 14- Memory bits M4 and M5 have been assigned to conveyor forward and reverse commands that control the conveyor manually.

4. Speed control

Please refer above VFD section 4.7 for better understanding.

5. Production report

As you can see I have used a rising pulse after X10, it will make sure a counter will receive single pulse during operation. Each pulse gets stored by date register D504.

I have assigned D504 to the numeric display on HMI. Thus, whenever a sensor X10 detects the presence of finished goods, the HMI will display the number of products.

5. REMOTE PLC PROGRAM ACCESS

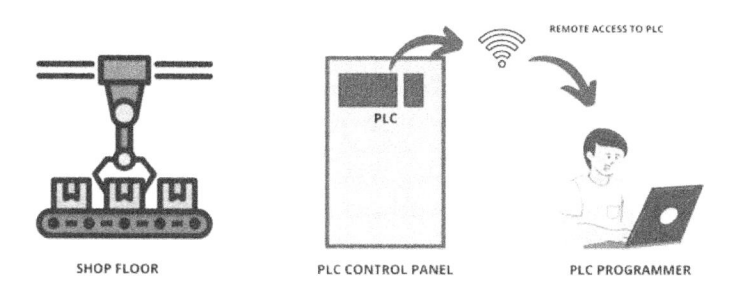

If a programmer wants to access a PLC program, it is necessary to connect a laptop to the PLC through ethernet, but for that, a programmer has to be available on shop floor. Nowadays, programmers use third party applications like Anydesk or teamviewer to get access. But in this case, it is necessary to have a connected laptop to the shop floor machine and both laptops should have installed software.

What if I tell you, there is another best way by which a programmer can get accessed into a program without connecting a laptop to the PLC. By using this method, you can connect your laptop from anywhere. In this chapter we are going to learn about the latest technology that helps programmers to get accessed into a PLC program remotely. Industrial machinery and equipment are built to last for a long period of time. After installation and commissioning and during the operation several breakdown problems can occur or changes to the PLC program can be needed to optimise and troubleshoot the PLC.

5.1 Needs for remote PLC control

Let's take a look at the different needs for remote PLC control.
1. Investigating causes of PLC problems

There are several PLC problems that can cause errors. Common PLC problems are communication errors, input/output module failures and power outage. A communication error can occur between the programmable computer and the devices it controls.

To investigate the potential PLC problem the use of PLC remote access helps to solve issues from any remote location. It provides a simple and secure method of connecting to a PLC controller which gives you full access and the ability to run for example a PLC scan. Remote access to PLCs, and connected devices, are great solutions for establishing remote support from any network-connected location. Managing a PLC with remote access can lead to quick configurations and deliver remote support to your industrial equipment.

2. PLC program changes

PLC programs are never final and due to changing customer requirements you will always need to make corrections and adaptations to the system. Even during operation, program changes are often necessary as the PLC program can almost always be improved after the first test run for optimisation.

All available offline and virtual PLC program testing tools can be used intensively to test program faults and run simulations. Corrections and modifications to the PLC program, transferring and testing of software can nowadays be easily done remotely to save valuable hours for the (service) engineer.

3. Remote commissioning of the entire system

PLC commissioning is a series of careful examinations done in a PLC control system to ensure that the controller, program, input and output devices conform to the specifications of the design. Once the final status of the PLC program is established, tested and the entire system commissioned, the system is ready to be handed over to the customer.

With the help of a local engineer or operator, remote commissioning can save on travelling expenses. Augmented reality solutions can support remote assistance.

5.2 Security to avoid unwanted access

By attaching a programmable logic controller to a network with Internet access, the device will be exposed to all of the same possible security threats as a computer.

According to a PLC access control security analysis, PLCs are a preferred target for industrial control system attackers. Therefore manufacturer provides several security measures to protect the machine builder and their customers with a secure Industrial IoT solution so they can troubleshoot PLC problems remotely without having to worry about security issues.

Using the advanced role based user system you are in control of each user's access rights. The device's built-in firewall blocks unwanted access to the PLC from outside the plant and separates the machine network from the corporate network.

For increased protection, a virtual private network (VPN) can be setup to increase the security by encrypting the data transmission when traveling over a public network - such as the Internet. Instead of opening all the ports that are needed to handle communication to the control network, one single authenticated network port passes the encrypted communication so the user can have all of the access as if they were inside the local area network.

5.3 How it works?

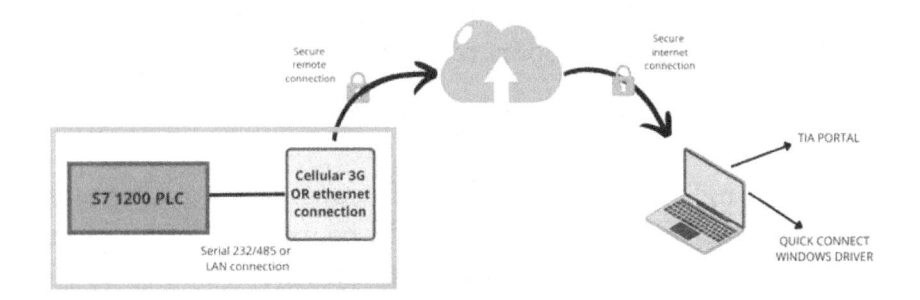

There are so many devices are available in the market that help to establish a communication between PLC and laptop through internet. In this section we are using Netbiter EC300 for better understanding.

On your office computer, you run the Net biter Quick Connect software and a secure tunnel is created to the Net biter EC350 which is connected to the remote PLC or machine. You can then simply open your regular configuration software (such as RS Logix or TIA Portal) and configure or debug just as if you were on site.

The connection is set up via the Net biter Argos data center in the cloud. You set up the connection in a matter of minutes after you have connected your Net biter gateway.

5.4 Remote access to CompactGuard Logix PLC via Ewon cosy 131

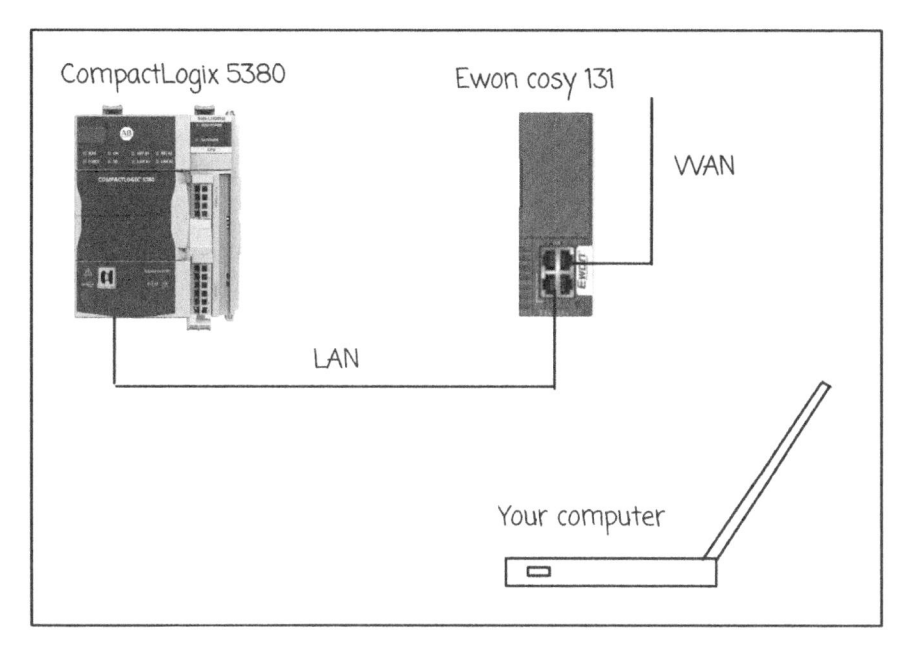

In this example, I am going to show how you can connect your PLC to ewon module and troubleshoot the PLC program. I have connected a compactLogix 5380 PLC to the ewon cosy 131 device.

The ewon manufacturer offers two software:

a. eBuddy - Configuration of ewon module

b. ecatcher- To establish a communication with PLC via ewon

A. Ewon cosy 131 module configuration using eBuddy

1. Create an account on TALK2M.

2 Connect your computer to the Ewon module via LAN cable.

3. Open eBuddy software, to see the connected module.

4. You can update firmware as well as change IP address of the module.

5. Open the browser and configure remaining settings.

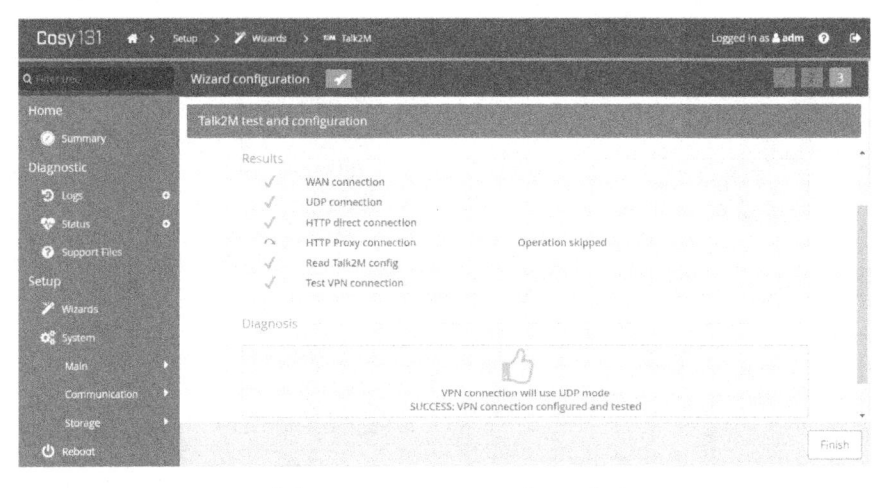

6. Connect WAN cable to WAN port of module.

B. Connect with compact GuardLogix 5380 controller using eCatcher

Now let's assume you have already configured the ewon module and connected it to the PLC.

1. After opening eCatcher software, you can see the connected ewon modules.

2. Select the module you wish to connect to.

3. Click on connect.

C. Configure the driver in RS Linx

a. The eCatcher software will automatically configure the IP address of eWon module.

b. Open RS Logix software and click on configure ethernet drivers.

c. You need to do mapping of the connected PLC.

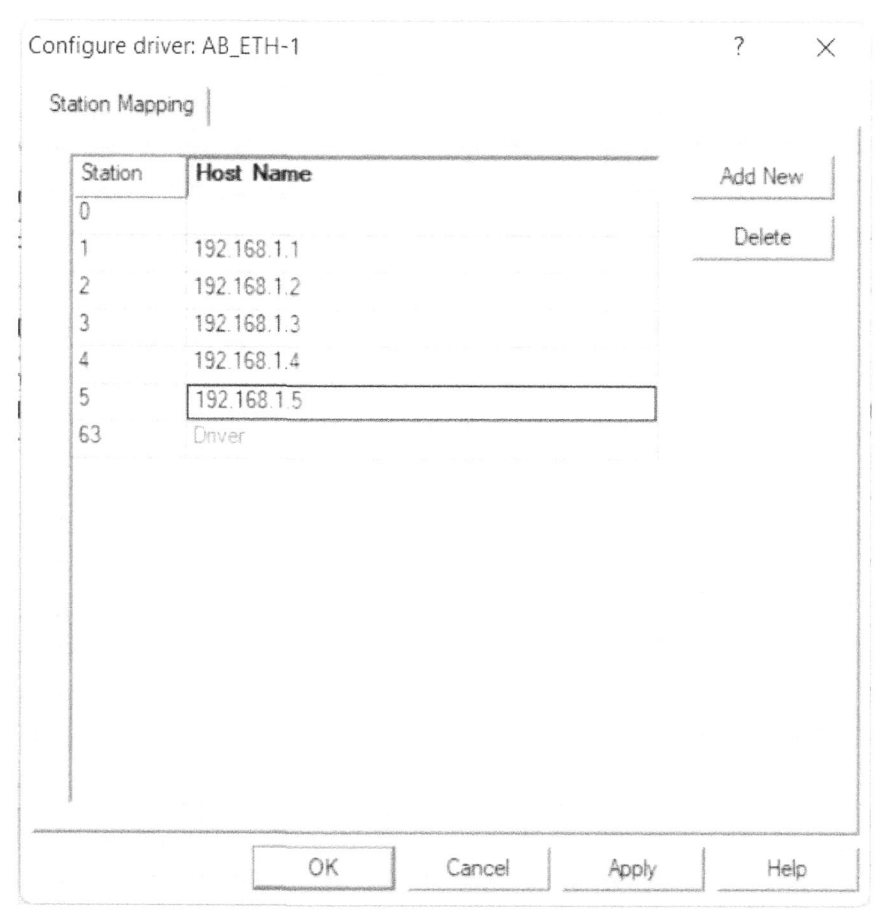

```
Configure driver: AB_ETH-1                           ?    ×

Station Mapping

  | Station | Host Name   |    | Add New |
  |---------|-------------|    |---------|
  | 0       |             |    | Delete  |
  | 1       | 192.168.1.1 |
  | 2       | 192.168.1.2 |
  | 3       | 192.168.1.3 |
  | 4       | 192.168.1.4 |
  | 5       | 192.168.1.5 |
  | 63      | Driver      |

         OK        Cancel       Apply       Help
```

AS you can see I have to map IP addresses of all the connected devices in the network.

d. Click on who active in the communication tap and check if all the devices are in the network.

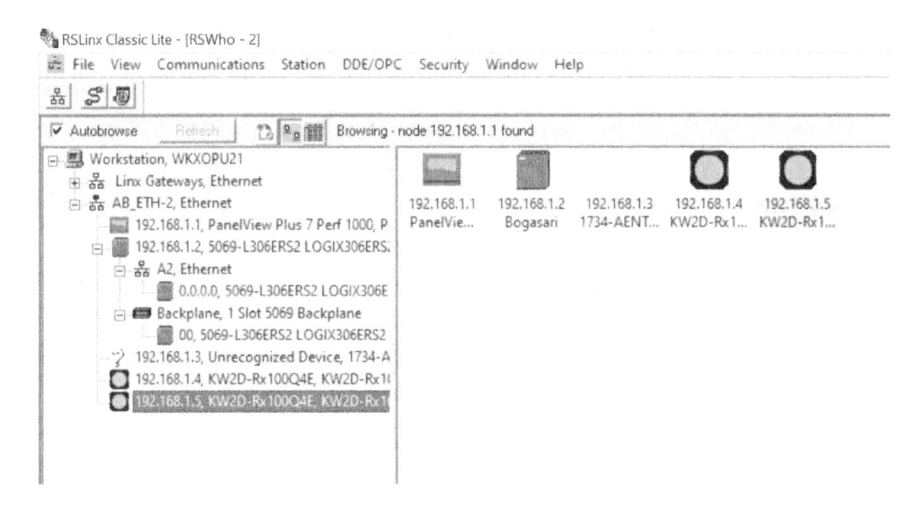

d. Go online using Studio 5000 Logix designer

Now open the PLC program in studio 5000 Logix designer and give a communication path to the software and connect with the PLC.

In this way you can establish a wireless communication with the PLC and access its program using ewon module from anywhere in the world without visiting the plant.

ASSIGNMENTS

Assignment 01
When the start PB is pressed light gets on for 10 seconds then gets off for 5 seconds. This cycle will continue until a stop PB is pressed.

Assignment 02
A pick and place mechanism consist of a pneumatic vacuum gripper, that picks up the pouches when a sensor detects the presence of a pouch and keeps that pouch in a box.
-A conveyor starts when the start PB is pressed. A pouch starts traveling over the conveyor.
-When a sensor detects the pouch, the conveyor stops after 2 sec. Then a slider moves towards the pouch, picks up the pouch, and keeps the pouch into the box.
-As soon as the gripper picks up the pouch, the conveyor starts after 3 sec.

Assignment 03
A Saw, Fan, and oil pump all go ON when a start button is pressed. If the saw has operated less than 20s, the oil pump should go off when the saw is turned off and the fan is to run for an additional 5s after the shutdown of the saw.
If the saw has operated for more than 20s, the fan should remain on until reset by a separate fan reset button and the oil pump should remain on for an additional 10 s after the saw is turned off.
Write a program that will implement this process.

Assignment 04
The oven is heated by an electrical heater, and inside there are ventilation motors to cool the oven after use. The electrical heater and the cooling fans should turn on simultaneously. The cooling fans have to turn on too, to circulate the hot air and spread the heat.
Since both the fans and the heater have to start at the same time, the two outputs should work simultaneously. But keep in

mind that the cooling fans have to run for some time after the heater is turned off.

Assignment 05

A classroom has a capacity of a maximum of 120 students. There are two doors, one for Entry and the other for Exit. When the number of students in the classroom is less than 120, the Entry door has a Greenlight on it which remains ON. When the number of students in the classroom is 120 or more than that, the Red light goes ON, turning OFF the Greenlight which indicates that the classroom has reached its maximum capacity and is full.

Problem Description

Considering the availability of two separate doors for Entry and Exit, two separate Proximity Switches can be used to detect the entry and exit of students. One proximity switch is mounted at the Entry door and the other at the Exit door.

Both the switches will generate two different outputs which can be then fed to PLC to operate the lights according to the Ladder Logic Program written in its memory. The counter must be used to count the number of students entering and exiting.

A comparator must also be used to compare the count value with the given maximum capacity of 120.

Assignment 06

Consider the design of a Burglar Alarm for a house. This alarm will be activated if an unauthorized person is detected by a Window Sensor or a Motion Detector. Implement this alarm System in PLC using Ladder Diagram programming language.

Assignment 07

In a water bottle manufacturing factory, different capacity bottles are manufactured from 1 liter to 20 liters. All bottles with range from 1 liter to 5 liters are collectively checked and moved to section 1 for separation. Similarly bottles from 6 liter to 10 liters moved to section 2 and bottles from 11 liter to 20 liters moved to section 3.

Check 1 liter to 5-liter bottles and switch on the green lamp.

Check 10 liter to 20-liter bottles and switch on the yellow lamp.
If a bottle is found above 20 liters, switch on the red lamp.

Assignment 08
When a start push button is pressed, the process will start. If the level of water in a reaction tank is low then start the pump till water level reaches high level. When it reaches a high level, stop the pump.
Now sense the PH value of liquid in the reaction tank, if the PH value is greater than 7, start the acid valve and let the acid enter into the tank. Simultaneously switch on the stirrer till the PH value becomes less than or equal to 4, then stop the stirrer and stop the acid valve. Then start the drain valve.
Stop the drain valve as soon as the liquid touches the low level of reaction tank and the inlet valve again.

Assignment 09
When the start PB is pressed both conveyors get on simultaneously. First conveyor gets off after 5 sec and the second conveyor gets off after 10 sec. Both conveyors remain off for 10 sec and this cycle continues until a stop button is pressed.

Assignment 10
A ladder logic for the forward and reverse direction of the motor and display pilot lights for the direction of the motor.

Assignment 11
A railway station has 3 platforms A, B and C. A train is coming into the station, it has to be given entry to platform A, if platform A is empty, otherwise to platform B. In case both platforms A and B are occupied, then it has to be given entry to platform C. If all platforms are occupied, then the train has to wait.

Assignment 12
When a switch is pressed, motor 1 gets started for 10 sec, then motor 2 starts for 8 sec, and motor 3 starts for 7 sec. This process repeats 3 times, then switches off all motors.

Assignment 13

A parking plot has a total capacity of Cars. A number of empty spots are displayed on the display outside the Parking Plot and which spots are available is to be indicated by LEDs. Implement this in PLC using Ladder Diagram programming language. The counter is used to count the number of empty spots.

Proximity Sensors or IR Sensors are used to detect the presence of the car. Here in this system, IR Sensors can be well installed to make this system cost-efficient since Proximity Sensors are costlier than IR Sensors.

The value of the counter is displayed on the display which is mounted outside the parking plot.

Assignment 14

When the entered password is OK (111111) and an enter key is pressed, the lock should get opened.

If the entered password is wrong and an enter key is pressed, the lock should not get opened and an alarm will turn on.

Assignment 15

A system consists of a single push button and two lamps.

-When a push button is pressed the first time, lamp1 will glow.

-When a push button is pressed a second time, lamp2 will glow.

-When a push button is pressed a third time, lamp1 will glow.

BOOKS IN THIS SERIES

Industrial Automation

Everything About Factory Automation

Looking to learn about the exiting world of automation? Look no further than this informative and friendly guide!

In this book, you'll explore amazing chapters on sensors, control panel, PLC with different PLC programming languages, HMI, VFD, and servo drive.

But don't worry this is not like a boring technical manual. I have written this book with a bunch of real world applications. Whether you're a curious beginner or an experienced technician looking to expand your knowledge you will enjoy this book.

Everything About PLC Programming

Are you interested in learning about Programmable Logic controller? Look no further than this friendly guide to PLC programming!

Whether you are a beginner or an experienced programmer, this book will take you on a journey through the fundamentals of PLC programming, including the use of industry standard software such as RS Logix 500, Siemens TIA, and GX Works 3.

With clear explanations and easy to follow examples, you'll learn how to design and implement PLC programs for variety of applications. From basic ladder logic to advanced function block programming. Additionally, you'll gain valuable insignt into troubleshooting and maintaining PLC systems, ensuring that your programs operate effectively. So why wait? Start your journey to PLC programming mastery today!

Advaced PLC Programming Part 1 using Studio 5000

In this comprehensive guide, you'll explore everything from basic of programming to advanced techniques and real world case studies. With a focus on studio 5000 Logix designer, you'll learn how to design and implement effective PLC programs. Whether you are a beginner or an experienced programmer, you'll find something new to discover in this book. From controlling VFD to programming HMI, this book covers it all. You'll gain insignt into important safety programming concepts and data collection techniques.

In addition to leaning aspect you will get explore to the real world examples with amazing case study.

Made in the USA
Coppell, TX
05 June 2024